Lecture Notes in
Computer Science

T0238225

Lecture Notes in Computer Science

Lecture Notes in Computer Science

Edited by G. Goos and J. Hartmanis

454

Volker Diekert

Combinatorics on Traces

 Springer-Verlag

Berlin Heidelberg New York London
Paris Tokyo Hong Kong Barcelona

Author

Volker Diekert
Institut für Informatik, Technische Universität München
Postfach 20 24 20, D-8000 München 2, FRG

CR Subject Classification (1987): F.m, F.3.2, F.4.2−3, G.2.1

ISBN 3-540-53031-2 Springer-Verlag Berlin Heidelberg New York
ISBN 0-387-53031-2 Springer-Verlag New York Berlin Heidelberg

Printing and binding: Druckhaus Beltz, Hemsbach/Bergstr.
2145/3140-543210 − Printed on acid-free paper

Für
Christine, Florian, Nikolai
und Michael

Foreword

Research on formal methods for the description and analysis of the behavior of nonsequential systems has increased considerably in the last decade. This has been mainly due to the rapidly growing importance of multiprocessor configurations, distributed systems and communication networks. Another important factor of acceleration was that a number of older lines of research in informatics, mathematics and logic appeared to fit together when seen under the new aspects of distributedness, concurrency and communication, and that the new theoretical results were greatly needed by practitioners in order to cope with the complexity of nonsequential systems.

There are two main approaches to a theory of nonsequential systems. The older one, created by C.A. Petri at the beginning of the 1960s, started out from the classical theory of sequential machines but uses as basic notions concurrency and asynchronous communication between distributed activities (which causes local state transformations). The fundamental concept is nonsequentiality, and sequential systems are simply the special cases where concurrency does not occur. The formal system model of this theory, the Petri net, was soon used widely in applications and inspired many theoretical investigations. The development of the Petri-net-related theory was particularly quick and fruitful when deeper connections to classical theories like those of formal languages, semilinear sets and vector addition systems were discovered and intensively used.

The second approach, developed at the end of the 1970s by R. Milner and C.A.R. Hoare, was more programming language oriented and started out from the ideas around λ-calculus; it is based on the synchronous communication of sequential processes and was the basis for several programming languages and influential theories for the specification and semantics of communicating sequential systems.

One of the reasons for the great success of the Milner-Hoare theory is that it stays very close to the traditional concept of sequential systems. This pertains not only to the structure (non-sequential systems are obtained as parallel compositions of sequential components) but also to the semantics: the behavior of a system is described from the point of view of one single sequential observer, who can only note sequences of actions or events. Thus concurrent activities of the system are noted in an arbitrary order, i.e. concurrency is modeled by nondeterminism. This allows us to use the large body of knowledge from formal language and automata theory and from logics and semantics of sequential programming languages. But it cannot cope adequately with all phenomena of nonsequential systems.

In Petri's approach a clear distinction between nondeterminism and concurrency is made: Concurrency is considered as a nontransitive relation, in particular it is different from simultaneity. Its complement is describing causal dependencies. Therefore the semantics is technically more complicated. Observations have to include the dependency of activities, i.e. are partially ordered structures for which not much classical theory previously existed. Although considerable progress has been made in this area, the mathematical treatment of partial order semantics in its full generality is quite cumbersome, such that it is worthwile to look for compromises between the two extremes.

The most successful idea for distinguisting concurrency from nondeterminism without departing too much from the well developed and nice sequential theory came from A. Mazurkiewicz as early as 1977. His idea simply is to endow the sequential observer of a nonsequential system with a small amount of information about the structure of the system, namely its concurrency relation. This, however, is only meaningful for unlabelled systems, where actions at different positions have different names, and all are observable. It soon turned out that this nice trick associates well-known mathematical objects with nonsequential systems: the free partially commutative monoids, which are indeed intermediate between the free (noncommutative) monoids used in the Milner-Hoare theory and the free (totally) commutative monoids occuring when Petri nets are viewed from the standpoint of sequential automata theory (namely as vector addition systems).

Following Mazurkiewicz's terminology, the theory of free partially commutative monoids is called *trace theory* when it is used for the semantics of nonsequential systems; these monoids are then called trace monoids and their elements traces since they denote traces of processes in nonsequential systems. Trace theory and other investigations of free partially commutative monoids have shown a vivid development over the last few years. It became clear that trace theory is mathematically deep and beautiful, semantically powerful and flexible, and a very challenging part of theoretical informatics. This was particularly obvious at the international workshop on free partially commutative monoids organized by V. Diekert as part of the activities of the ESPRIT Basic Research Action No. 3166 "Algebraic and Syntactic Methods in Computer Science (ASMICS)" in October 1989 in Kochel am See, Bavaria, see [Die90c] and [Die90d].

V. Diekert has contributed considerably to this development in three different ways, which are brought together in this LNCS volume. He presents and extends the algebraic and combinatorial foundations of trace theory in a coherent way and embeds it in or relates it to other mathematical or informatical theories. He develops a new theory of replacement systems especially for trace monoids, and he enlarges the range of application of the theory to a much wider class of Petri nets than that considered previously.

This text is self-contained apart from basic theoretical informatics and is written in an easily readable form such that it may serve as a textbook for a first-year graduate course.

Munich, April 1990   Wilfried Brauer

Preface

These notes are a treatise on partially commutative words, commonly called traces. The underlying algebraic structures are free partially commutative monoids which have been introduced by P. Cartier and D. Foata in order to solve some combinatorial problems of rearrangement. In computer science they have been recognized as an algebraic model for concurrency. This is essentially due to the work of A. Mazurkiewicz who also introduced the basic notions and the word *trace*. Mazurkiewicz used traces as a partial order semantics for safe Petri nets. Since his initiating work a systematic study of traces under various aspects has begun.

The present volume is an extended and revised version of the *Habilitationsschrift* of the author written at the Technical University of Munich. It contains some basic material on traces including Ochmanski's characterization of recognizable trace languages and Zielonka's theory of asynchronous automata. The third chapter is devoted to an application of this theory to a modular approach for the computation of Petri net languages. This is based on so-called local morphisms between nets which yield certain homomorphisms between the corresponding trace monoids and allows a convenient treatment of the synchronization of nets. Another part of these notes concern the Möbius function and its relations to semi-Thue systems. In the last chapter we generalize the concept of semi-Thue systems to a combinatorial theory of rewriting on traces. This can be viewed as an abstract calculus for transformations of concurrent processes.

Acknowledgements

These notes would have not been written without Professor W. Brauer. I wish to thank him for constant encouragement and support over the last few years. Many fruitful discussions with him stimulated the progress of my work and led to substantial improvements.

Many thanks also to Professors M. Paul and D. Perrin who were further referees of my *Habilitationsschrift*.

I am also greatly indebted to all members of the group *Theoretische Grundlagen der Informatik* at Hamburg and Munich. In particular, I am grateful to Heino Carstensen, Jörg Desel, Matthias Jantzen, Dirk Hauschildt, Astrid Kiehn, Manfred Kunde, Klaus-Jörn Lange, Oliver Schoett, Dirk Taubner and Walter Vogler for various exchanges of ideas.

This work has also benefited very much from discussions with my colleagues: Ron Book, Christian Choffrut, Robert Cori, Christine Duboc, Zoltán Ésik, Dominique Foata, Massimiliano Goldwurm, Hendrik-Jan Hoogeboom, Roman König, Giancarlo Mauri, Ernst Mayr, Antoni Mazurkiewicz, Yves Métivier, Maurice Nivat, Edward Ochmanski, Friedrich Otto, Dominique Perrin, Jean-Eric Pin, Nicoletta Sabadini, Jacques Sakarovitch, Volker Strehl, Klaus Reinhardt, Grzegorz Rozenberg, Wolfgang Thomas, Celia Wrathall and Wieslaw Zielonka.

Many of them are gathered in the EBRA-working group No. 3166 "Algebraic and Syntactic Methods in Computer Science (ASMICS)" and/or in the EBRA-project No. 3148 "Design Methods Based on Nets (DEMON)". I thank these ESPRIT Basic Research Actions, ASMICS and DEMON, for all kinds of support which helped me to complete this work.

Last but not least I thank Harald Hadwiger for his excellent work in transforming the handwritten manuscript into LaTeX and I thank Springer-Verlag for accepting it into its Lecture Notes series.

Munich, April 1990   Volker Diekert

Contents

Introduction

Parallelism or concurrency is one of the fundamental concepts in computer science. One motivation for its use is simply to increase the efficiency of sequential programs. With the help of massive parallelism we may hope to solve problems which otherwise would be intractable. Besides this, parallelism serves also as a basis for a general methodology for designing or specifying software (and hardware).

Another important aspect of parallelism is that many existing and an increasing number of forthcoming systems are inherently concurrent. The wide range of such systems includes distributed, decentralised, loosely or tightly coupled, embedded, and reactive systems. The design, the analysis and the verification of such concurrent systems usually turns out to be a very difficult task.

There are essentially two reasons for this. The first one is that the behavior of concurrent systems is much more complex than that of comparable sequential centralized systems. The presence of concurrency implies that in most cases the behavior of a component cannot be understood in isolation. One has always to take into account its multiple interactions with other simultaneously active parts of the system. The second reason concerns the formal methods. They are far from being sufficiently developed for a rigorous formal treatment of concurrent systems. But without a satisfactory theoretical basis and a good formal framework, analysis and verification of concurrent systems can only be based on ad hoc proofs. Since finding ad hoc proofs is difficult this is usually expensive. Even worse, often it leads to mistakes. With the increasing importance of concurrency in computer science this situation is unacceptable. Therefore we need a deeper mathematical theory of parallelism.

A common approach to a systematic study of parallelism requires as a first step that one can associate with each concurrent system a well-defined mathematical object which denotes important aspects of its behavior. On a certain level of abstraction we may say that the setting of a concurrent system is given by a set of atomic actions $X = \{a, b, c, \ldots\}$ together with a specification of which actions can be performed independently or concurrently. Such a specification is given in our approach by an independence relation $I \subseteq X \times X$. For technical reasons we shall assume that I is irreflexive and symmetric. This reflects that no action can act concurrently to itself and that independence is mutual. Actions which are not independent are called dependent. A concurrent process in this abstraction is a labelled acyclic graph where the labels of nodes are actions and edges represent an ordering between dependent actions. No edges are drawn between independent actions. This is called the dependence graph, or simply the trace, of the process. The trace describes a concurrent process just as a sequential process is described by a string. Simple as it is, the success of trace theory owes very much to this observation.

An execution of a concurrent process has to respect only the ordering between the dependent actions. It follows that if u, v are processes and a, b are independent actions, then $uabv$ and $ubav$ are nothing but different sequential runs of the same concurrent process, or different sequential observations of the same concurrent run.

This leads us to the equivalence relation generated by these pairs $uabv, ubav$, i.e., to the equivalence generated by the rearrangement of independent actions. Since this equivalence is a congruence, it defines a quotient monoid of X^* which is called the free partially commutative monoid associated with X and I. These monoids have been introduced into combinatorics by P. Cartier and D. Foata. In our model, concurrent processes are associated via their dependence graphs with elements of a free partially commutative monoid. It turns out that the correspondence between dependence graphs and elements of a free partially commutative monoid is one-to-one.

From the graphical representation of a concurrent process by its trace, we may read immediate information. For example, we can see the effect of unbounded parallelism on the reduction in computation time. Namely, in a first parallel step we may perform the minimal elements of the dependence graph. In a second parallel step we may perform the minimal elements of the remaining graph, and so on. (The parallel steps are exactly the factors in the so-called Foata normal form; their number is just the minimal parallel computation time.)

In order to obtain more insight into traces, consider a concurrent process p and let a, b be (not necessarily different) dependent actions which occur in p. Then there is a directed path in the trace of p where all vertices with label a or b occur. Disregarding all other letters, we obtain a string over $\{a, b\}^*$. A basic result in trace theory states that a trace is uniquely determined if we know these strings for all dependent $a, b \in X$. The connection between the dependence graphs and these strings over dependent letters has been established, apparently for the first time, by R. Keller in [Kel73]. In his paper the connection was shown directly without involving free partially commutative monoids explicitly. Also, the semantics of the Foata normal form as maximal parallelism is shown there.

However, it is mainly due to of the work of A. Mazurkiewicz that the attention of computer scientists was focused on free partially commutative monoids. Mazurkiewicz also introduced the notion of a *trace*. He was led to it when he studied the behavior of safe Petri nets, which are a widely adopted model for concurrent systems. A state of the concurrent system is represented by a distribution of tokens over the places and the actions of the system are the transitions of the net. This model incorporates intrinsically an independence relation between different transitions: they are independent if they are not adjacent to a common place. It was discovered by Mazurkiewicz that the behavior of a safe Petri net is best described by traces [Maz77]. Since the initiating work of Mazurkiewicz traces have been taken up by many people and a profound study under various aspects has begun. The reader may also consult [Maz87], [AR88] and [Per89] for recent excellent overviews.

Let us mention some of the areas where trace theory has been applied. M. Flé and G. Roucairol used traces as a tool for serializability problems in distributed data bases, see [FR82].

The theory of COSY-path-expressions, [LTS79], can be directly formulated in terms of trace theory. There is also a close connection between abstract programming languages, like Milner's calculus of communicating systems *CCS*, [Mil80], or Hoare's language *CSP* of communicating sequential processes, [Hoa85], and traces in the sense used here – although this is less direct than in COSY.

The theory of formal languages over traces started with the investigation of recognizable subsets of free partially commutative monoids in [Fli74] . Recognizable sets are fundamental for any formal language theory. They are of further importance since they correspond to finite-state controlled structures. Therefore a lot of attention has been paid to them; in particular the relation to rational sets and decidability questions were investigated, see [AW86], [AH87], [BBMS81], [Bra76], [CM85], [CP85], [Mét86], [Sak87]. (This list is not meant to be complete.) Highlights in this area are Ochmanski's characterization of recognizable languages, [Och85] and Zielonka's work on asynchronous automata, [Zie87]. Of great interest too is the theory of solving equations over partially commutative variables, see [Dub86].

The application of trace theory to Petri nets has been of constant interest. In particular, the synchronization of trace languages as a tool for a modular computation of net languages has been recognized, see [Maz87].

Only recently have connections between trace theory and rewriting systems been studied. There are several approaches. The link to the theory of semi-Thue systems is the basis for [BL87], [Ott87], [NO88], or [Wra88]. In [Ott89] term rewriting methods from [Jou83] are applied to traces. In [Die87a] we initiated a theory which uses trace rewriting directly. This work has been continued in [Die89] and [Die90a]. Here, it is summarized and extended in the last chapter.

Some other, but not all, of the material above will be treated in this monograph. The contents are organized in detail as follows: The purpose of the first chapter is to introduce the general concepts. We start with an informal example in Sect. 1.1. In Sect. 1.2 we develop the combinatorial background. In particular we give a graph theoretical characterization for the factorization of a trace into *subtraces*. This elementary operation allows to transform graphical and formal proofs into each other. Section 1.4 gives a categorical approach to the general embedding theorem. The basis is some sort of duality between dependence graphs and free partially commutative monoids. This can be expressed by a contravariant functor. It is one of the reasons that we base our notation on dependence graphs instead of independence graphs. An important well-known result states that a covering of dependence graphs corresponds to an embedding of free partially commutative monoids. In particular, it is possible to represent a trace as a tuple of words. In Sect. 1.5 we present some efficient algorithms for computations which follow such a representation. These algorithms are suitable for implementation and up to constant factors they are of optimal time complexity. In particular we present a linear time algorithm for a subtrace test.

Chapter 2 is devoted to the relation between recognizable and rational sets. It is the least original chapter. No really new result is stated, but we have rewritten proofs. We present a straight path to Ochmanski's theorem which characterizes recognizable languages in terms of so-called recognizable expressions, and we have included Zielonka's theory of asynchronous automata.

The behavior of Petri nets in terms of traces is investigated in Chap. 3. In the literature this is limited to the case of safe nets, since for this class of nets, traces give a complete picture of the possible concurrency. For higher nets this is not true anymore. Nevertheless it is our aim to show that traces are a useful tool for Petri nets in general. We introduce so-called *local morphisms* between nets. The term 'local' is derived from the fact that the defining property can be checked locally at the places of the nets. The image net of a local morphism is always transition bounded; the local morphism itself may merge independent transitions together, but also places may be merged. Using local morphism we give a categorical definition of the synchronization of nets as a push-out. This generalizes the usual definition which is restricted to safe nets and to the case of place disjoint nets.

The rough idea behind all this is to imitate some concepts developed in topology. Following these concepts the notion of a covering is introduced. In its simplest case this corresponds to a covering by transition bounded subnets, or more precisely to the natural mapping from the disjoint union of nets to its synchronization. The hope is therefore that if we have a covering of a net N by a net N' then the behavior of N' has a simpler description than the behavior of N. Now, local morphisms relate the behavior of nets to each other. But, as in Sect. 1.4, we have a relation in the opposite direction. (However, this is quite natural. Think of the inclusion of nets, where one may restrict the language of the larger net to a smaller one. This is also an exchange of directions.) The main result of Sect. 3.1 states that if $N' \to N$ is a covering of nets then the trace language of N may be identified with the intersection of the trace language of N'

and the image of the free partially commutative monoid which is given by the transitions of N together with their natural independence relation. Our theorem depends heavily on the use of traces. It could not even be formulated adequately without them. In fact, the importance of the theory of traces for higher nets is not the descriptive power of traces, but the convenient way we may compute with them.

In its simplest case the theorem yields back the result that the trace language of the synchronization of two nets is the synchronization of the two trace languages.

The synchronization of trace languages is studied in detail in Sect. 3.3. This section is based on a joint work with W. Vogler. The main result gives a graph-theoretical as well as an algebraic characterization of when the synchronization can be computed really locally on the components. This settles an open problem of [AR86]. Our characterization also has a close connection to the problem of reconstructability, as it is considered in [CM85].

In Sect. 3.4 we show that our graph theoretical condition is *NP*-complete. Hence, for large dependence alphabets it might be very difficult to decide whether the synchronization of trace languages allows a local description. However, in restricted cases we are able to give polynomial time algorithms.

The last two chapters concern the investigation of replacement systems. In Chap. 4 we concentrate on the classical theory of semi-Thue systems. In Sect. 4.1 we first recall the concepts of this theory and we introduce minimal critical pairs. We show that the confluence of noetherian systems can be checked on these minimal critical pairs. This result is important since it may increase the efficiency of the Knuth-Bendix completion procedure for semi-Thue systems.

In Sect. 4.2 we show that every (finitely generated) free partially commutative monoid has a presentation by some (finite) complete semi-Thue system. This complete system is constructed over the set of 'parallel steps'. It yields thereby another and elegant proof for the Foata normal form theorem.

Due to a result of F. Otto, [Ott87], no complete presentation exists over a minimal set of generators, in general. In Sect. 4.3 we relate the result of Otto to a well-known conjecture on Möbius functions for free partially commutative monoids. In fact, we settle the conjecture. We thereby establish a direct link between the theory of semi-Thue systems and formal power series. This seems to be very promising for future research. A first step is taken in Sect. 4.4 where we consider arbitrary noetherian semi-Thue systems. We compute the formal inverse of the characteristic series over the irreducible words in terms of left-hand sides. This leads to the notion of an *overlapping-chain*. Short overlapping-chains have a direct interpretation and it turns out that the overlapping-chains of length three exactly correspond to minimal critical pairs, which were mentioned above. In fact, it was the search for an interpretation of overlapping-chains that led us to the consideration of minimal critical pairs.

In the final chapter we work with replacement systems directly over traces. These replacement systems generalize and unify the notion of semi-Thue systems and vector replacement systems. We base the replacement upon the factorization of traces, more precisely upon the rewriting of subtraces. The general idea of trace replacement systems, however is, to provide an abstract calculus for transformations of concurrent processes.

Let us explain this in more detail. Suppose, we are given a specification by a set of equations over concurrent processes. This may be viewed as a system of rules over traces. The question arises whether two concurrent processes may be transformed into each other by successive applications of the specified rules. In an algebraic translation this question becomes a word problem of a trace replacement system. In general, such a word problem is undecidable. It is solvable if the corresponding system is finite, noetherian and confluent.

The general procedure to find such systems is described by the Knuth-Bendix completion. But in working with traces new difficulties and phenomena arise. These concern mainly the critical pairs which are of overall importance in the completion procedure. We show that even for one-rule systems infinite sets of critical pairs may occur. We therefore have to put restrictions on the systems to ensure finite sets of critical pairs. The restrictions we give are decidable and always satisfied in the case of semi-Thue systems or vector replacement systems. But they go beyond this. We also propose a slightly modified version of the Knuth-Bendix completion which takes care of the restrictions only once in a final step. We give examples where we obtain complete systems which satisfy the restrictions even though at the beginning of the procedure they were violated. We also present examples of complete trace replacement systems which may not be decoded as any complete semi-Thue system or vector replacement system. Thus, our theory is a real generalization. It may be applied to situations where classical approaches fail.

In dealing with trace replacement systems, complexity questions are of central interest. For a given finite system one reduction step can be computed in linear time. This yields a time complexity for computing irreducible normal forms which is a square in the derivation length. We do not know whether this square time bound is optimal, but the restriction to semi-Thue systems or vector replacement systems yields linear time, due to a result of R. Book, [Boo82].

In virtue of this phenomenon we state a decidable sufficient condition which ensures that for a given finite noetherian system, irreducible normal forms can be computed in time linear in the derivation length by a very simple algorithm. This algorithm is a straightforward generalization of the one for semi-Thue systems. In a slightly different form the same algorithm has also been used for traces by C. Wrathall in [Wra88] and by R. Book and H. Liu in [BL87]. The main new result is that our condition is strong enough to decide the confluence in this case.

We also present a uniform algorithm to compute irreducible normal forms which works for every finite noetherian system. Its implementation depends heavily on the construction of Zielonka's asynchronous automata. If the condition mentioned above is satisfied then this algorithm succeeds in linear time, but in general its worst case behavior is square time. It is a challenging open problem whether a different approach may yield substanially better results.

In the last section we restrict our attention to systems where all left hand sides have some geometric interpretation as *cones* or *blocks*. Cones and blocks are certain types of traces. The simplest examples of cones are words in a free monoid and of blocks are vectors in a free commutative monoid.

Thus, already these systems generalize the classical theory of semi-Thue or vector replacement systems. Nevertheless with respect to the problem of deciding confluence and to the time complexity question these systems behave as well as we are used to in the classical cases.

Theses notes are written on an elementary level; i.e., no particular familiarity is required except for some basic mathematical background. We give complete proofs in most cases.

After having read the first chapter, the reader may go straight to one of the following chapters.

1. Free Partially Commutative Monoids

1.1 An Introductory Example

To give some impression of the background of trace theory we start with a rather informal example. This example reflects only a very small aspect of trace theory, but it may give a first flavour.

Consider a distributed database consisting of a set of records and a set of update transactions which may be performed concurrently. In order to avoid incorrect updates let us follow a protocol which allows concurrent read and exclusive write. To fix ideas, we suppose that our database has records w, x, y, z containing some integer values, and we intend to proceed a sequence of atomic transactions from the the following list:

(a) $x := x + y$

(b) $x := w + x$

(c) $y := y + z$

(d) $w := 2y + z$

(e) $z := y + 2z$

(f) $x := w + x + y$

(g) $x := w + x + 2y$

From the variables involved we obtain a dependence relation between the transactions which expresses that they may not be executed concurrently. We represent the dependence structure by a graph where the vertices are the transactions and edges are between dependent ones. In our example this yields Figure 1.1.

Let p be any sequence of transactions. Clearly, if we can write $p = uxyv$ for some independent (i.e., not dependent) transactions x, y then the sequential execution of $p' = uyxv$ yields the same result as p. Since 'yielding the same result' is, a priori, an equivalence, we may consider the equivalence relation generated by these pairs $(uxyv, uyxv)$ where x, y are independent. Every such equivalence class is called a trace. If p, p' belong to the same trace then the effect of p on the database is the same as p'.

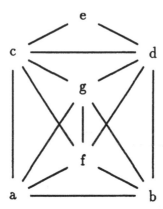

Figure 1.1. Dependence graph of transactions

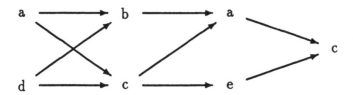

Figure 1.2. The Hasse diagram of the trace $p = dabcaec$

Since the equivalence defined above is in fact a congruence, the set of traces form in a natural way a monoid. Thus we can compute with traces as we are used to do with sequences.

Moreover, traces provide us with some sort of partial order semantics. To define it, start with a sequence $p = x_1 \ldots x_n$ of transactions. With p we associate a directed acyclic labelled graph as follows:

We take a vertex set $\{v_1, \ldots, v_n\}$, where each vertex v_i is labelled with the value of x_i for $i = 1, \ldots, n$. Then we draw directed edges from v_i to v_j if $i < j$ and if x_i and x_j are dependent. As usual when working with partial orders we may omit redundant edges by drawing the Hasse diagram only.

For example, for $p = dabcaec$ we obtain the graph as in Figure 1.2. The overall effect of p on the database is $w := 2y+z$, $x := w+x+2y+z$, $y := 2y+4z$, and $z := y+3z$ and it is clear that this overall effect is the same for every sequence of transactions which yields the same graph as above.

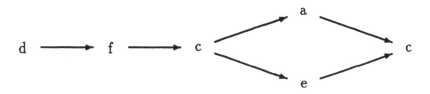

Figure 1.3. The Hasse diagram of p where ab is replaced by f

This graph is called the dependence graph of p. In fact, the trace belonging to the sequence p can be identified with its dependence graph. From this graph we get immediately information on what can be done in parallel: In a first step we may execute exactly those transactions which are minimal in the dependence graph. Taking away the corresponding elements, we obtain new minimal elements which can be performed in a second parallel step and so forth.

This greedy algorithm yields the minimal parallel execution time. In the example above we obtain a sequence of four parallel steps:

$$\{a, d\}, \{b, c\}, \{a, e\}, \{c\}.$$

Later we shall see that this corresponds to the Foata normal form of the trace.

The assumption that update transactions are atomic is a strong simplification. More realistically, we should view a transaction as a finite sequence of atomic actions. This leads to a finer trace model. In particular, it is suitable to handle problems of serializibility, see e.g. [FR85].

Let us briefly discuss another point. In our introductory example ab has the same effect as the transaction $f = (x := w + x + y)$. It follows that in every sequence where ab occurs, we may replace ab by f. This is obviously the case in $p = dabcaec$, but we may replace ab by f also in $p' = adcbeac$, simply because p and p' describe the same trace. The replacement of ab by f may be viewed as an application of a transformation rule. It may be defined on the dependence graph by substituting the subgraph $a \to b$ by f and drawing edges in the 'right-way'. For p it yields a dependence graph as in Figure 1.3. Observe that the transaction f has to be executed before c and after d, otherwise we would not obtain the same result in the execution anymore. Since we have replaced two actions by one, the sequential time of the replaced process has decreased. But note that the parallel time needs one step more.

A further inspection of the transactions shows that an execution of aba has the same effect as $g = (x := x + 2y + w)$. But in the graph belonging to the trace $p = dabcaec$, a replacement of the subgraph $a \to b \to a$ by g would lead to serious problems. Indeed, the effect of $a\{b,c\}a$ on the record x is $(x := x + 2y + z + w)$, the effect of gc it is $(x := x + 2y + w)$, and of cg it is $(x := x + 2y + 2z + w)$. Thus $gc \neq a\{b,c\}a \neq cg$. Since c occurs after the first a and before the second a the effect of the execution aba depends on the context c. Algebraically this means that aba is not a factor of the trace in Figure 1.2. Taking this into account we shall develop an abstract calculus of transformations which is based on a replacement of so-called subtraces in the last chapter.

1.2 Basic Definitions

In this section we introduce the basic notions of trace theory and we state some elementary results which will be used throughout. A *dependence alphabet* is a pair (X, D) where X is a (finite) alphabet and $D \subseteq X \times X$ is a reflexive, symmetric relation, the *dependence relation*. The complement of D is irreflexive and symmetric; it is called the *independence relation* and is denoted by $I = X \times X \backslash D$. Let $=_c$ be the equivalence relation on X^* which is generated by all pairs of the form $(uabv, ubav)$ for $u, v \in X^*$, $(a, b) \in I$. Then $=_c$ is a congruence and the set of congruence classes forms a monoid $M(X, D)$ with the multiplication $[u] \cdot [v] = [uv]$. The neutral element is the class of the empty word, it is denoted by 1.

These monoids are called *free partially commutative*, [Lal79]. They were first studied in mathematics by P. Cartier and D. Foata, [CF69], in order to solve some combinatorial problems. In computer science A. Mazurkiewicz [Maz77] introduced them in connection with the analysis of safe net systems.

Mazurkiewicz called an element of a free partially commutative monoid a *trace*. This notation is now standard and adopted here. Also we will use the notation *trace monoid* as an abbreviation for *finitely generated free partially commutative monoid*.

To explain the term *free partially commutative monoid* consider any other monoid M. Then a homomorphism of $M(X, D)$ to M is completely determined by a mapping $f : X \to M$ (freeness) such that $f(a) \cdot f(b) = f(b) \cdot f(a)$ for all $(a, b) \in I$ (partial commutativity).

For example, let (X', D'), (X, D) be dependence alphabets such that $X' \subseteq X$ and $D' \subseteq D$. Consider the mapping $p : X \to M(X', D')$ given by $p(x) = x$ if $x \in X'$ and $p(x) = 1$ otherwise. Obviously, it holds $p(a)p(b) = p(b)p(a)$ in $M(X', D')$ for all $(a, b) \in I$. Thus, we obtain a homomorphism $p : M(X, D) \to M(X', D')$. We refer to this mapping as the *canonical projection* of $M(X, D)$ onto $M(X', D')$.

We have two extreme cases of dependence relations, namely $D = X \times X$ and $D = \mathrm{id}_X$. The first one yields the free monoid $M(X, X \times X) = X^*$, the second one the free commutative monoid $M(X, \mathrm{id}_X) = \mathbb{N}^{|X|}$ with basis X. By $\mathbb{N}^{|X|}$ we denote the set of mappings from X to the natural numbers $\mathbb{N}$ which are zero almost everywhere. The monoid structure is given by the usual addition. If X is finite then $\mathbb{N}^{|X|} = \mathbb{N}^X =$ set of all mappings from X to $\mathbb{N}$. Since $\mathrm{id}_X \subseteq D \subseteq X \times X$ for every dependence relation we always have the following sequence of canonical projections:

$$X^* \to M(X, D) \to \mathbb{N}^{|X|}$$

The composition of these mappings coincides with the usual Parikh mapping from words to vectors.

Let $w \in X^*$ be a word and let $t \in M(X, D)$ be the congruence class of w. We shall write $t = [w]$ to denote this, or simply $t = w$, if there is no risk of confusion. For the word w we have the usual terms of *length*, denoted by $|w|$, for a letter $a \in X$ the *a-length* of w, denoted by $|w|_a$ (= number of a occuring in the string w), and *alphabet* of w, denoted by $\mathrm{alph}(w)$ (= $\{a \in X \mid |w|_a \geq 1\}$). The meaning of these terms translates immediately to traces. Thus we may write $|t| := |w|$, $|t|_a := |w|_a$, $\mathrm{alph}(t) := \mathrm{alph}(w)$ for $t = [w]$. It is also very useful to extend the independence relation $I = X \times X \setminus D$ to words over X and traces in $M(X, D)$. We say that $u, v \in X^*$ ($u, v \in M(X, D)$ respectively) are *independent* if $\mathrm{alph}(u) \times \mathrm{alph}(v) \subseteq I$. Frequently we shall simply write $(u, v) \in I$ for this, too.

So far, traces are equivalence classes of words and we can use words to represent them. This allows ambiguity since different words may denote the same trace. This ambiguity vanishes if we represent traces in some normal forms. Historically the first one was given by D. Foata.

In order to define it, let $\mathcal{F}$ be the following set of finite non-empty subsets of pairwise independent letters:

$$\mathcal{F} = \{F \subseteq X \mid \emptyset \neq F \text{ is finite and } (a, b) \in I \text{ for all } a, b \in F, a \neq b\}.$$

Each $F \in \mathcal{F}$ is called an *elementary step* and it yields a trace $[F] \in M(X, D)$ by taking the product over its elements, $[F] = \prod_{a \in F} a$. Note that this is well-defined since $F \in \mathcal{F}$ consists of finitely many pairwise commuting elements, only.

With the definition of elementary steps the Foata normal form theorem can be stated as follows.

Theorem 1.2.1 ([CF69, Thm 1.2]) *Let $t \in M(X, D)$ be a trace. There exists exactly one sequence of elementary steps $(F_1, \ldots, F_r)$, $r \geq 0$, $F_i \in \mathcal{F}$ for $1 \leq i \leq r$ such that $t = [F_1] \ldots [F_r]$ and for all $b \in F_i$, $2 \leq i \leq r$ there is some $a \in F_{i-1}$ with $(a, b) \in D$.* □

There are several different proofs of this basic theorem. In the last chapter we will present a proof which is based on complete semi-Thue systems. It is

also possible to derive the Foata normal form from Corollary 1.4.7 below, see
Remark 1.4.9 in Sect. 1.4. As a direct consequence from Theorem 1.2.1 free
partially commutative monoids are cancellative, see [CF69, Cor. 1.3].

A second possibility to get unique representations of traces is given by their
so-called lexicographical normal forms:

Let $t \in M(X, D)$ be a trace then the number of strings $w \in X^*$ with $t = [w]$ is
finite and these strings must have same length. If there is a linear order on X then
among these $w \in X^*$ with $t = [w]$ there is a string which is lexicographically the
first one. This string is called the *lexicographic normal form* of t. The following
proposition characterizes lexicographic normal forms. This proposition is due to
A. V. Anisimov and E. Knuth, it has been published in [Per84, Thm. 2.2.].

Proposition 1.2.2 *Let (X, D) be a dependence alphabet and $\leq$ be a linear ordering of X. Then, a word $w \in X^*$ is the lexicographic normal form of a trace
in $M(X, D)$ if and only if for each factor aub of w with $a, b \in X$, $u \in X^*$,
$(au, b) \in I$ it holds $a < b$.*

Proof: Clearly, if a lexicographic normal form w of a trace contains a factor
aub with $(au, b) \in I$ then $a < b$. For the other direction assume that $w, w' \in X^*$
denote the same trace and that w is lexicographically before w'. Then $w =
w_1 a w_2$, $w' = w_1 b w_2'$ for some $w_1, w_2, w_2' \in X^*$ and $a < b$. Since free partially
commutative monoids are cancellative, aw_2 and bw_2' denote the same trace. But
then $bw_2' = buaw_2''$ for some $u, w_2'' \in X^*$, $a \notin \mathrm{alph}(u)$, and $(bu, a) \in I$. $\square$

Corollary 1.2.3 *The set of lexicographic normal forms is a regular language.* $\square$

Besides the string representation of traces, traces are given by certain graphs.
This graphical representation is extremely helpful.

By a *labelled acyclic graph* (over X) we understand a triple (V, E, λ) where
(V, E) is a finite directed acyclic graph with vertex set V and edge set $E \subseteq V \times V$
and where $\lambda : V \to X$ is a labelling function.

We say that a labelled acyclic graph (V, E, λ) is a dependence graph (over
(X, D)) if edges are between different vertices with dependent labels and no
further edges exist. The set of isomorphism classes of dependence graphs over
(X, D) form a monoid with the multiplication

$$[V_1, E_1, \lambda_1][V_2, E_2, \lambda_2] \;=\; [V_1 \dot\cup V_2, E_1 \dot\cup E_2 \dot\cup \{(x, y) \in V_1 \times V_2 \mid
(\lambda(x), \lambda(y)) \in D\}, \lambda_1 \dot\cup \lambda_2]$$

and the neutral element $1 = [\emptyset, \emptyset, \emptyset]$.

For a moment, let $\mathrm{Dep}(X, D)$ be this monoid of isomorphism classes. We
have a natural mapping $dep : X \to \mathrm{Dep}(X, D)$ which associates with each letter
$a \in X$ the dependence graph consisting of one point with label a.

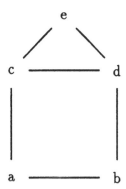

Figure 1.4. Our home-example of a dependence alphabet (X, D)

If $(a, b) \in I$ then $\mathrm{dep}(a)\,\mathrm{dep}(b)$ and $\mathrm{dep}(b)\,\mathrm{dep}(a)$ yield the same graph which consists of two points, labelled a and b, and without any edge. Thus, as explained above, there is a natural homomorphism $\mathrm{dep} : M(X, D) \to \mathrm{Dep}(X, D)$. The mapping dep is surjective since $\mathrm{Dep}(X, D)$ is generated by its one-point graphs.

To see that dep is injective let $t = [F_1] \ldots [F_r]$ be a trace in Foata normal form. Clearly, $\mathrm{dep}(t) = \mathrm{dep}([F_1]) \ldots \mathrm{dep}([F_r])$ and one easily sees that for $1 \le i \le r$ we have $a \in F_i$ if and only if there is a vertex x in $\mathrm{dep}(t)$ with label a and the longest directed path ending in x has length $i - 1$. Hence, from $\mathrm{dep}(t)$ we get back the Foata normal form of t.

The following well-known proposition is therefore nothing but an easy translation of Foata's normal form theorem.

Proposition 1.2.4 *The free partially commutative monoid $M(X, D)$ is canonically isomorphic to $\mathrm{Dep}(X, D)$.* $\square$

In the following, a trace $t \in M(X, D)$ will simply be identified with its dependence graph $\mathrm{dep}(t)$. If the dependence alphabet (X, D) is fixed then the whole information of t is contained in the Hasse diagram of $\mathrm{dep}(t)$. Thus, we need not to draw redundant edges. On the other hand, sometimes it is more useful to think of a trace as a labelled partially ordered set. We obtain this partial order as the transitive closure of the directed edges. Again, no information is lost. We therefore have several possibilities to represent a trace and we shall always choose the most appropriate one.

Example: Let (X, D) be given by Figure 1.4 where edges are drawn between different dependent letters.

Let $t \in M(X, D)$ be the trace which is given by the string $eacbd$. Then this trace is also given in the following way:

a) Foata normal form: $\{a, e\}, \{b, c\}, \{d\}$,

b) lexicographic normal form: *abecd*

c) dependence graph:

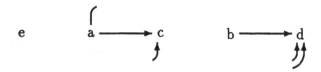

d) Hasse diagram:

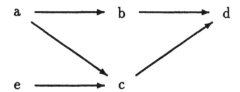

e) the labelled partial order which is obtained from the dependence graph or
the Hasse diagram above by setting $x \leq y$ if there is a directed path from
vertex x to vertex y. $\square$

Minimal and maximal elements of traces will be used frequently, for example,
in inductive proofs. Let $t \in M(X, D)$ be a trace. The set of minimal (maximal
respecitively) elements of t is defined by $\min(t) = \{a \in X \mid t = [au]$ for some
$u \in X^*\}$ $(\max(t) = \{a \in X \mid t = [va]$ for some $v \in X^*\})$. Since these elements
commute, they define a trace by taking their product. We denote again this
trace by $\min(t)$ ($\max(t)$ respectively). Then for some $u, v \in M(X, D)$ we have
$t = \min(t)u = v\max(t)$. Of course, the sets $\min(t)$ and $\max(t)$ respectively
coincide with the labels of the minimal respectively maximal elements of the
dependences graph of t. The trace $\min(t)$ is also exactly the first factor in the
Foata normal form of t. On the other hand, $\max(t)$ need not to be the last factor.
This results since Foata normal forms are computed from the left to the right.
For example, let $t = aab$ where a and b are independent. Then the Foata normal
form of t is $\{a, b\}\{a\}$, but $\max(t) = \{a, b\}$.

More generally, let $t \in M(X, D)$ be a trace and let $l \subseteq t$ be any subset in the
dependence graph of t. Then the subgraph induced by l is a dependence graph.
Hence it defines a unique trace $l \in M(X, D)$.

We shall use the following convention. If $u \subseteq t$, $v \subseteq t$ are subsets of t then $t = u \cup v$, $(t = u\dot{\cup}v)$, means that the labelled vertex set of t is the (disjoint) union of the labelled vertex sets of u and v. The (disjoint) union is not meant with respect to the edge sets. We say that $l \subseteq t$ is a *subtrace* if for some $u, v \subseteq t$ with $t = u\dot{\cup}l\dot{\cup}v$ we may write $t = ulv$. Thus, a subtrace corresponds to a factor of t. The other way round, if we have a factorization $t = ulv$ for some $u, l, v \in M(X, D)$ then there is a unique way to identify these elements with subtraces of t.

Viewing a trace t as a labelled partial order, we have the following simple but important characterization of subtraces:

Lemma 1.2.5 *Let $t \in M(X, D)$ be a trace and $l \subseteq t$ be a subset of the dependence graph of t. Let $\leq$ be the induced partial order on t. Then $l \subseteq t$ is a subtrace of t if and only if for all $x, z \in l$, $y \in t$ with $x \leq y \leq z$ it holds $y \in l$.*

Proof: Let $l \subseteq t$ be a subtrace and write $t = ulv$ for some subtraces $u, v \subseteq t$ with $t = u\dot{\cup}l\dot{\cup}v$. Let $x, z \in l$ and $y \in t$ with $x \leq y \leq z$. Then $x \leq y$ implies $y \notin u$ and $y \leq z$ implies $y \notin v$, hence $y \in l$.

Assume now, that t is a trace and $l \subseteq t$ is a subset such that for all $x, z \in l$, $y \in t$ with $x \leq y \leq z$ it holds $y \in l$. If $\min(t) \subseteq l$ and $\max(t) \subseteq l$ then we have $t \subseteq l$, hence $l = t$ and l is a subtrace. In the other case say without restriction, there is $a \in \min(t)$ with $a \notin l$. Then we may write $t = at'$ with $l \subseteq t'$. By induction on the cardinality of $t \setminus l$ we obtain that l is a subtrace of t'. Hence $t' = u'lv$ and $t = au'lv$. $\square$

Remark 1.2.6 To test whether $l \subseteq t$ defines a subtrace we used the labelled partial order given by t. The reason is that for the dependence graph of t or its Hasse diagram such a test is 'global'. For these graphs it is not possible to decide in the direct neighborhood of l whether $l \subseteq t$ is a subtrace: Let (X, D) be given by:

$$a \quad \text{------} \quad b \quad \text{------} \quad c \quad \text{------} \quad d \quad \text{------} \quad e.$$

Consider a situation as in Figure 1.5. In the first case $\{a, e\} \subseteq t_1$ is a subtrace, whereas $\{a, e\}$ is no subtrace of t_2. $\square$

From the characterization of subtraces above we obtain the following fact:

Remark 1.2.7 If $t \in M(X, D)$ is a trace and $l_1 \subseteq t$, $l_2 \subseteq t$ are subtraces then $l_1 \cap l_2$ is subtrace of t, too. $\square$

$t_1 = ab^n d^n e =$

$t_2 = ab^n cd^n e =$

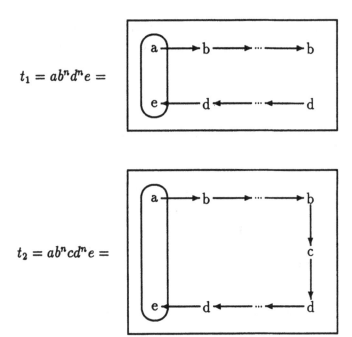

Figure 1.5. The set $l = ae$ is a subtrace of t_1 but not of t_2

If $t \in M(X, D)$ is a trace and $l \subseteq t$ is a subtrace, then we define the subtrace $\text{pre}(l)$ of elements *before* l, $\text{suf}(l)$ of elements *behind* l, and $\text{ind}(l)$ of elements *independent of* l by the following formulae:

$$
\begin{aligned}
\text{pre}(l) &= \{x \in t \setminus l \mid \exists y \in l : x \leq y\}, \\
\text{suf}(l) &= \{z \in t \setminus l \mid \exists y \in l : y \leq z\}, \\
\text{ind}(l) &= t \setminus (\text{pre}(l) \cup l \cup \text{suf}(l)).
\end{aligned}
$$

Note that ind defines a symmetric relation on subtraces l_1, l_2 of t. We have $l_1 \subseteq \text{ind}(l_2)$ if and only if $l_2 \subseteq \text{ind}(l_1)$. Furthermore, if $l_1 \subseteq \text{ind}(l_2)$ then we have $(l_1, l_2) \in I$. A picture for a partitition of traces based on these subtraces is given by Figure 1.6. The ind(l)-part has been devided into two parts in order to reflect the factorization of the next proposition.

Proposition 1.2.8 *Let $u, v, t \in M(X, D)$ be traces and $l \subseteq t$ be a subtrace of t. Then we have $t = ulv$ if and only if $u = \text{pre}(l)u_1$, $v = v_1 \text{suf}(l)$ for some $u_1, v_1 \in M(X, D)$ with $u_1 v_1 = \text{ind}(l)$.*

Proof: Of course, $t = \text{pre}(l)u_1 l v_1 \text{suf}(l)$ holds for all factorizations $u_1 v_1 = \text{ind}(l)$. Hence, one direction is trivial. For the other direction, assume $t = ulv$ and identify $u \subseteq t$, $v \subseteq t$ with subtraces. We must have $\text{pre}(l) \subseteq u$, $\text{suf}(l) \subseteq v$; and we can write $u = \text{pre}(l)u_1$, $v = v_1 \text{suf}(l)$ for some $u_1, v_1 \in M(X, D)$. Since $u \cap \text{suf}(l) = \emptyset$, we have $u_1 \subseteq \text{ind}(l)$ and hence $t = \text{pre}(l)u_1 l v_1 \text{suf}(l) = \text{pre}(l)lu_1 v_1 \text{suf}(l)$. Since $t = \text{pre}(l)l \, \text{ind}(l) \, \text{suf}(l)$ by the other direction, we obtain $u_1 v_1 = \text{ind}(l)$ by the cancellativity of free partially commutative monoids. $\square$

Let $l \subseteq t$ be any subset of a trace t. Then we may define the *generated subtrace* of l in t by the set $\{y \in t \mid \exists x, z \in l : x \leq y \leq z\}$. This set is a subtrace of t and will be denoted by $\langle l \rangle$. An alternative definition of $\langle l \rangle$ is given by the intersection of all subtraces $l' \subseteq t$ which contain l. Thus, $\langle l \rangle$ is the smallest subtrace of t which contains the subset l of t.

1.3 The Levi Lemma for Traces

Levi's Lemma for strings has the following generalization:

Proposition 1.3.1 ([CP85, Prop 1.3]) *Let $x, y, z, t \in M(X, D)$ be traces. Then the following assertions are equivalent:*

i) $xy = zt$

ii) $x = ru$, $y = vs$, $z = rv$, $t = us$ for some $r, u, v, s \in M(X, D)$ such that $(u, v) \in I$.

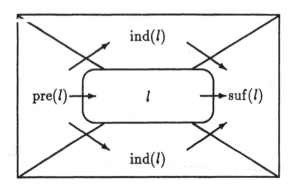

Figure 1.6. Picture of a trace t with a subtrace $l \subseteq t$.

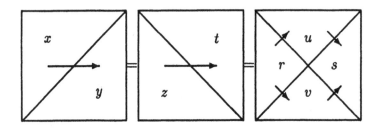

Figure 1.7. The "proof" of Levi's Lemma for traces

*Furthermore, if $xy = zt$ then the corresponding traces $r, u, v, s \in M(X, D)$ in ii)
are uniquely determined and for all letters $a \in X$ we have: $|r|_a = \min(|x|_a, |z|_a)$,
$|u|_a = |x|_a - |r|_a$, $|s|_a = \min(|y|_a, |t|_a)$, and $|v|_a = |y|_a - |s|_a$.*

Proof: ii) $\Rightarrow$ i): trivial, since $(u, v) \in I$ implies $uv = vu$.

i) $\Rightarrow$ ii): Let $w = xy = zt$ and identify x, y, z, t with subtraces of w. Define
the subtraces $r = x \cap z$, $u = x \cap t$, $v = y \cap z$, $s = y \cap t$ as in Figure 1.7. Directly
from the definition, we obtain $x = ru$, $y = vs$, $z = rv$ and $t = us$. We have to
show $(u, v) \in I$, only. Since $w = rvus = ruvs$, we must have $\mathrm{pre}(v) \subseteq r$ and
$\mathrm{suf}(v) \subseteq s$, see Proposition 1.2.8. Hence $u \subseteq \mathrm{ind}(v)$ since r, u, v, s are pairwise
disjoint subtraces. But now $u \subseteq \mathrm{ind}(v)$ implies $(u, v) \in I$. Thus, we have ii). To
see that r, u, v, s are uniquely determined, simply observe that $(u, v) \in I$ implies
$\mathrm{alph}(u) \cap \mathrm{alph}(v) = \emptyset$. Hence we obtain the formulae $|r|_a = \min(|x|_a, |z|_a)$
and $|s|_a = \min(|y|_a, |t|_a)$ for all $a \in X$. The other two formulae follow. The
uniqueness of these subtraces is also clear. $\square$

Let M be any monoid and $s, t \in M$ be elements. We say that s is a *proper
divisor* of t if $t = usv$ for some $u, v \in M$ with $u \neq 1$ or $v \neq 1$. We say that the

divisor relation is noetherian if there are no infinite chains $t_0, t_1, t_2, \ldots$ such that t_{i+1} is a proper divisor of t_i for all $i \geq 0$. The following fact is well-known.

Lemma 1.3.2 *Let M be a monoid where the divisor relation is noetherian. Then M is generated by the set $X = (M \setminus \{1\}) \setminus (M \setminus \{1\})^2$. The inclusion $X \hookrightarrow M$ induces a surjective homomorphism $\pi : X^* \to M$ such that $\pi^{-1}(1) = \{1\}$.*

Proof: The fact that $\pi : X^* \to M$ is surjective follows easily by noetherian induction. To see that $\pi^{-1}(1) = \{1\}$ assume the contrary and let $w \in X^+$ be the shortest non-empty word such that $\pi(w) = 1$. By definition of X the word w is not a letter. Thus, we may write $w = uv$ with $\pi(u) \neq 1$ and $\pi(v) \neq 1$. It follows $1 = \pi(w) = \pi(u)\pi(v)$ and 1 is a proper divisor of itself. This contradics the hypothesis that the divisor relation is noetherian. $\square$

Remark 1.3.3 The divisor relation is noetherian in free partially commutative monoids. This follows since traces have a length and the length of a trace t is zero only if $t = 1$. The surjective homomorphism mentioned in the lemma above is the canonical projection $p : X^* \to M(X, D)$. Elementary examples where the divisor relation is not noetherian are non-trivial groups or the boolean semi-ring $\mathbf{B} = \{0, 1\}$.

Definition: *Let M be a monoid, we say that Levi's Lemma holds in M if for all $x, y, z, t \in M$ which satisfy $xy = zt$ there are $r, u, v, s \in M$ such that $x = ru$, $y = vs$, $z = rv$, $t = us$ and $uv = vu$.*

Remark 1.3.4 Levi Lemma holds in free partially commutative monoids by Proposition 1.3.1. But it holds also in every abelian group or in the boolean semi-ring $\mathbf{B}$. In particular, Levi Lemma alone is not enough to characterize free partially commutative monoids. $\square$

The following algebraic characteriziation of free partially commutative monoids is due to Christine Duboc. It is a generalization of Levi's result on free monoids, [Lev44]. Later we shall use the corollary below in order to prove that certain fibered products are free partially commutative.

Theorem 1.3.5 ([Dub86, Thm 3.1.6]) *A monoid is free partially commutative if and only if the divisor relation is noetherian and Levi's Lemma holds in it.*

Proof: By the remarks above only the 'if' part of the assertion requires a proof. Let M be a monoid such that the divisor relation is noetherian and the Levi Lemma holds in M. Define $X = (M \setminus \{1\}) \setminus (M \setminus \{1\})^2$, and let $p : X^* \to M$ be the surjective homomorphism from Lemma 1.3.2. Say that $a, b \in X$, $a \neq b$ are dependent if $ab \neq ba$ in M. This gives us a dependence alphabet (X, D), and it is clear that p factorizes through $M(X, D)$. Thus, we obtain a natural surjective homomorphism, denoted again $p : M(X, D) \to M$. We prove by noetherian induction that p is injective. Let $t_1, t_2 \in M(X, D)$ such that $p(t_1) = p(t_2)$ and assume that every proper divisor of $p(t_1)$ has exactly one inverse image. If $t_1 \in (X \cup \{1\})$ then $t_1 = t_2$ since $p^{-1}(x) = \{x\}$ for all $x \in (X \cup \{1\})$. If $t_1 \notin (X \cup \{1\})$ then we may write $t_1 = xy$, $t_2 = zt$ with $x, z \in X$ and $y \neq 1$, $t \neq 1$. Since $p(x)p(y) = p(z)p(t)$ and p is surjective, Levi's Lemma for M tells us that there are traces $r, u, v, s \in M(X, D)$ such that $p(x) = p(r)p(u)$, $p(y) = p(v)p(s)$, $p(z) = p(r)p(v)$, $p(t) = p(u)p(s)$ and $p(uv) = p(vu)$. Since $p(x)$, $p(y)$, $p(z)$ and $p(t)$ are proper divisors of $p(t_1)$ we have $x = ru$, $y = vs$, $z = rv$, and $t = us$. Now, if $r \neq 1$ or $s \neq 1$ then $p(uv)$ is a proper divisor of $p(t_1)$, too; and it follows $uv = vu$ in $M(X, D)$. Hence $t_1 = ruvs = rvus = t_2$ in this case. Therefore we are reduced to show $t_1 = t_2$ for $r = s = 1$. If $r = s = 1$ then $u = x \in X$ and $v = z \in X$; hence $u, v \in X$ are independent letters and $t_1 = uv = vu = t_2$ are the same trace by the very definition of (X, D). □

Corollary 1.3.6 *A submonoid of a free partially commutative monoid is free partially commutative if and only if Levi's Lemma holds in it.*

Proof: Let M' be a submonoid of a monoid M such that the divisor relation is noetherian in M. Then the divisor relation is noetherian in M', too. Hence the corollary follows directly from Theorem 1.3.5. □

Remark 1.3.7 The statement of Theorem 1.3.5 given here is slightly different from [Dub86]. C. Duboc does not use the property that the divisor relation is noetherian. Instead she demands that the monoid has a length, i.e., a homomorphism to $\mathbb{N}$ such that the inverse image of 0 is the identity. Clearly, if a monoid has a length then the divisor relation is noetherian. But the converse is false: Consider $M = \{a, b, c\}^*/abc = cb$. Since the congruence classes are finite, the divisor relation is noetherian; but M has no length. Thus, our condition is a little bit weaker. □

1.4 A Categorial Approach to the General Embedding Theorem

One of the fundamental properties of free partially commutative monoids states that every such monoid is embeddable in a direct product of free monoids. We will give a proof for this fact from a categorial viewpoint. In fact, this approach yields more general results and it clearifies the real connection between free partially commutative monoids and dependence alphabets. In the following we use the language of categories. For the notions the reader may consult the textbook of S. MacLane [Mac70] or any other books on categories. Nevertheless, the main results can be understood without any knowledge in category theory by the explicit definitions given below. Some results of this section may also be found in [Fis86], where the ideas have been developed in collaboration with the present author. A previous version of this section has also been presented in [DV88].

Every dependence alphabet (X, D) may be viewed as an undirected graph where edges are drawn between different dependent letters. Conversely, given any undirected graph $G = (X, E)$ with vertex set X and edge set $E \subseteq \binom{X}{2} = \{\{x, y\} \mid x, y \in X, x \neq y\}$ then G corresponds to the dependence alphabet (X, D) with $D = \{(x, y) \in X \times X \mid \{x, y\} \in E \text{ or } x = y\}$; let us denote the associated free partially commutative monoid with $M(G) = M(X, D)$. In order to give a functorial meaning to M, we have to say what we understand by a morphism $h : (X', E') \to (X, E)$ of undirected graphs. We define it as a mapping $h : X' \to X$ such that $h^{-1}(x)$ is finite for all $x \in X$ and such that $\{x', y'\} \in E'$ implies $\{h(x'), h(y')\} \in E$. Note that in particular we have $h(x') \neq h(y')$ for all edges $\{x', y'\} \in E'$. The assumption that the fibers $h^{-1}(x)$ are finite is made only to include infinitely generated free partially commutative monoids in our theory. This assumption is not necessary if one deals with finite alphabets only. Clearly, undirected graphs form a category with this kind of morphisms.

Let $h : G' \to G$ be a morphism of undirected graphs, $G' = (X', E')$, $G = (X, E)$ then we define a homomorphism of monoids in the opposite direction $M(h) : M(G) \to M(G')$ by $M(h)(y) := \prod_{x \in h^{-1}(y)} x$ (=product over all elements in $h^{-1}(y)$) for $y \in X$ and we extend $M(h)$ by multiplication $M(h)([y_1 \ldots y_n]) = [M(h)(y_1) \ldots M(h)(y_n)]$. To see that $M(h)$ is well-defined let $x_1, x_2 \in X'$ and $y_1 = h(x_1)$, $y_2 = h(x_2) \in X$. Assume that there is no edge between y_1 and y_2. Then there is no edge between x_1 and x_2 neither, hence $x_1 x_2 = x_2 x_1$ in $M(G')$. For $y = y_1 = y_2$ this means that the finite product $\prod_{x \in h^{-1}(y)} x$ is a well-defined trace, for $y_1 \neq y_2$ this means that the images of independent letters commute; hence $M(h)$ is a well-defined homomorphism.

Proposition 1.4.1 *The construction above defines a contravariant functor M from the category of undirected graphs to the category of free partially commutative monoids.*

Proof: Obviously, $M(\mathrm{id}_G) = \mathrm{id}_{M(G)}$ for the identity mapping. Let $f : G \to$

G', $g : G' \to G''$ be morphisms of undirected graphs. Let z be a vertex of G'' then $(gh)^{-1}(z)$ is the disjoint union $\bigcup\limits_{y \in g^{-1}(z)} h^{-1}(z)$. Hence $M(gh)(z) =$

$$\prod\limits_{y \in g^{-1}(z)} M(h)(z) = M(h)(\prod\limits_{y \in g^{-1}(z)} y) = M(h)M(g)(z). \;\; \square$$

Example: i) Let A, B be any sets and $f : A \to B$ be a mapping such that $f^{-1}(b)$ is finite for all $b \in B$. We may view f as morphism of undirected graphs $f : (A, \emptyset) \to (B, \emptyset)$. The associated free partially commutative monoids are $\mathbb{N}^{|A|}$ and $\mathbb{N}^{|B|}$ respectively. The homomorphism $M(f)$ has a very natural interpretation in this case:

$$M(f) = f^* : \mathbb{N}^{|B|} \to \mathbb{N}^{|A|}, \qquad m \mapsto m\, f : A \xrightarrow{f} B \xrightarrow{m} \mathbb{N}.$$

ii) Let (X', D'), (X, D) be dependence alphabets such that $X' \subseteq X$ and $D' \subseteq D$. The functor M applied to the following diagram of inclusions:

$$
\begin{array}{ccccc}
(X, X \times X) & \hookleftarrow & (X, D) & \hookleftarrow & (X, \mathrm{id}_X) \\
\uparrow & & \uparrow & & \uparrow \\
(X', X' \times X') & \hookleftarrow & (X', D') & \hookleftarrow & (X, \mathrm{id}_{X'})
\end{array}
$$

yields the following diagram of canonical projections:

$$
\begin{array}{ccccc}
X^* & \to & M(X, D) & \to & \mathbb{N}^{|X|} \\
\downarrow & & \downarrow & & \downarrow \\
X'^* & \to & M(X', D') & \to & \mathbb{N}^{|X'|} \quad \square
\end{array}
$$

For undirected graphs G_1, G_2 we denote their disjoint union by $G_1 \dot\cup G_2$. The complex product $G_1 * G_2$ is defined by the disjoint union together with the additional edges $\{x_1, x_2\}$ for all vertices x_i of G_i, $i = 1, 2$. Obviously the functor M behaves on these constructions as follows:

Proposition 1.4.2 *Let G_1, G_2 be undirected graphs and $M_i = M(G_i)$ be the associated free partially commutative monoids, $i = 1, 2$. Then the functor M applied to the canonical injections:*

$$G_i \; \hookrightarrow \; \underset{\text{(disjoint union)}}{G_1 \dot\cup G_2} \; \hookrightarrow \; \underset{\text{(complex product)}}{G_1 * G_2,}$$

yields the canonical projections:

$$\underset{\text{(free product)}}{M_1 * M_2} \; \to \; \underset{\text{(direct product)}}{M_1 \times M_2} \; \to \; M_i \;\; , i = 1, 2. \;\; \square$$

Another very elementary fact is established next.

Lemma 1.4.3 *Let $h : G' \to G$ be a morphism of undirected graphs. Then for all vertices x' of G' and all traces $t \in M(G)$ we have $x' \in \text{alph}(M(h)(t))$ if and only if $h(x') \in \text{alph}(t)$.*

Proof: Trivial □

The main interest in the contravariant functor results from the next theorem. This theorem was stated first in [Fis86, Thm. 2.1].

Theorem 1.4.4 *Let $h : G' \to G$ be a morphism of undirected graphs and $M(h) :$ $M(G) \to M(G')$ be the associated homomorphism of free partially commutative monoids. Then it holds:*

i) *The homomorphism $M(h)$ is surjective if and only if h is injective.*

ii) *The homomorphism $M(h)$ is injective if and only if h is surjective on vertices and edges.*

Proof: Let $G' = (X', E')$, $G = (X, E)$ and $M' = M(G')$, $M = M(G)$. For simplification we denote the homomorphism $M(h)$ by $h^* : M \to M'$.

i): Let $h : G' \to G$ be injective. We may view this as an inclusion, $X' \subseteq X$, $E' \subseteq E$ and the morphism $h^* : M \to M'$ becomes the canonical projection of $M((X, E))$ onto $M((X', E'))$.

Now, let $h^* : M \to M'$ be surjective and $x, y \in X'$ such that $h(x) = h(y)$. We have to show $x = y$. Since h^* is surjective we find a trace $t \in M$ with $h^*(t) = x$. Hence $alph(h^*(t)) = \{x\}$. Applying Lemma 1.4.3 twice we have $h(y) = h(x) \in \text{alph}(t)$ and $y \in \text{alph}(h^*(t)) = \{x\}$; hence $x = y$.

ii): Assume first that $h^* : M \to M'$ is injective. If h would be not surjective on vertices then there would be some $y \in X$ such that $h^{-1}(y)$ is empty. This implies $h^*(y) = 1$ which is impossible.

If h would be not surjective on edges then there would be some edge $\{y, z\} \in E$ such that all letters in $h^{-1}(\{y, z\})$ commute in M'. This implies $h^*(yz) = h^*(zy)$ in M' which is impossible for injective h^* since $yz \neq zy$ in M.

The most interesting part of the theorem is the other direction of ii) which will be proved now by contradiction using functorial properties. Let $h^*(t) = h^*(z)$ for some $t, z \in M(G)$ with $t \neq z$. Take $u \in M(G)$ of maximal length such that $t = ut_1$, $z = uz_1$ for some t_1, z_1. Then we may assume that we can write $t_1 = at_2$ for some $a \in X$, $t_2 \in M(G)$. Consider the graph inclusion $i_a : (\{a\}, \emptyset) \hookrightarrow (X, E)$. Since h is surjective on vertices we have $i_a = hf$ for some $f : (a, \emptyset) \hookrightarrow (X', E')$. Hence $i_a^* = f^* h^*$ by functorial properties. Since

$h^*(t) = h^*(z)$ it follows $|t|_a = i_a^*(t) = i_a^*(z) = |z|_a$. Hence $z_1 = vaw$ for some $v, w \in M(G)$. We may assume that v has minimal length for such a factorization. In particular $a \notin v$. On the other hand by the maximality of u we may conclude that v contains a letter $b \in X$ which depends on a. Hence we find an edge $e = \{a, b\} \in E$. Now consider the graph inclusion $i_{a,b} : (\{a, b\}, e) \hookrightarrow (X, E)$. We apply the same argument again. Since h is surjective on edges we have $i_{a,b} = fh$ for some $f : (\{a, b\}, e) \hookrightarrow (X, E)$. It follows $p_{a,b}(t) = i_{a,b}^*(t) = i_{a,b}^*(z) = p_{a,b}^*(z)$ where $p_{a,b} : M(G) \to \{a, b\}^*$ denotes the canonical projection. Since $\{a, b\}^*$ is cancellative we have $p_{a,b}(u) = p_{a,b}(v)$. But $p_{a,b}(u)$ starts with the letter a while $p_{a,b}(u)$ starts with the letter b. This is a contradiction. $\square$

Corollary 1.4.5 (General Embedding Theorem)

Let G be an undirected graph and let $\{G_j \mid j \in J\}$ be a family of subgraphs.

For $j \in J$ let $\pi_j : M(G) \to M(G_j)$ denote the canonical projection and let $\pi : M(G) \to \prod_{j \in J} M(G_j)$ be the homomorphism to the direct product given by $\pi(t) = (\pi_j(t))_{j \in J}$. Then the canonical mapping π is injective if and only if $G = \bigcup_{j \in J} G_j$.

Proof: Since the alphabet of each trace is finite, it is easy to see that we are reduced to prove the corollary for finite index sets J only. In the case of finite J we have a canonical morphism of undirected graphs from the disjoint union $\dot{\bigcup}_{j \in J} G_j$ to G. Hence a diagram:

$$G_j \longrightarrow G \quad = \quad \bigcup_{j \in J} G_j$$

$$\dot{\bigcup}_{j \in J} G_j \quad (= \text{disjoint union})$$

We may apply the functor M to this diagram and we obtain:

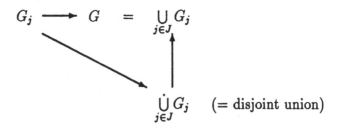

$$M(G_j) \xleftarrow{\ \pi_j\ } M(G)$$

$$\xleftarrow{p_j} \qquad \downarrow \pi$$

$$\prod_{j \in J} M(G_j)$$

In this diagram p_j denotes the j-th projection, π_j and π are as above. Thus the result follows directly from Theorem 1.4.4. □

Remark 1.4.6 Note that the direct product $\prod_{j \in J} M(G_j)$ is, in general, not a free partially commutative monoid since if $\{M_j \mid j \in J\}$ is any family of non-trivial free partially commutative monoids then $\prod_{j \in J} M_j$ is free partially commutative if and only if J is finite.

Indeed, if J is finite then $\prod_{j \in J} M_j$ is such a monoid and if J is infinite then it is enough to observe that the divisor relation is not noetherian for the infinite direct product $\prod_{j \in J} M_j$ where $M_j = \mathbb{N}$ for all $j \in J$.

How does this fit in our categorial set up? Let $G = \bigcup_{j \in J} G_j$ and let $\tilde{G} = \dot{\bigcup}_{j \in J} G_j$ be the disjoint union of the subgraphs G_j, $j \in J$. Then $M(\tilde{G})$ is the *weak product* $\bigoplus_{j \in J} M(G_j)$. This means it is the submonoid of the direct product which consists of those elements where almost all components are 1. (The plus sign $\oplus$ is used here to denote weak products. Note however that if differs from the categorial direct sum of monoids. This is the free product which is denoted by the sign $*$.) Now, we obtain a natural morphism $\tilde{G} \to G$ in our category of undirected graphs only if for each vertex x of G there are at most finitely many $j \in J$ such that $x \in G_j$. In this case, of course, we obtain a natural embedding $\pi : M(G) \hookrightarrow \bigoplus_{j \in J} M(G_j)$. But in general the weak product is not large enough to allow such an embedding. We have to use the direct product. This is unavoidable. It can be shown that the "infinite starfish" $M = M_0 * (\bigoplus_{i>0} M_i)$ with $M_i \cong \mathbb{N}$ for all $i \geq 0$ is not embeddable in any weak product of free monoids. □

A graph is called *complete*, or a *clique*, if any two vertices are connected by an edge. Since every graph is covered by cliques and cliques correspond to free monoids, a special case of Theorem 1.4.4 is the following embedding theorem. For trace monoids it was stated in [CL85, Lemma 3.1] and [CP85, Prop. 1.1]; at least implicitly it also stated in [Kel73, Lemma 2.5]. In terms of histories it can be derived from [Shi79].

Corollary 1.4.7 (Embedding Theorem) *Every free partially commutative monoid is a submonoid of a direct product of free monoids.* □

For each graph there is one natural choice for a covering by cliques. We can take its edge set and its isolated vertices. One obtains the following version of the embedding theorem:

Corollary 1.4.8 *Let (X, D) be a dependence alphabet, $E = \{\{x, y\} \mid (x, y) \in D\}$ and $Z = \{z \in X \mid (y, z) \notin D \; \forall y \in X, y \neq z\}$. Then the canonical projections define an embedding:*

$$\pi : M(X, D) \hookrightarrow (\prod_{\{x,y\} \in E} \{x, y\}^*) \times (\bigoplus_{z \in Z} z^*) \qquad \square$$

Remark 1.4.9 Let us show how to derive the Foata normal form theorem, Theorem 1.2.1, from the corollary above. For a trace $t \in M(X, D)$ the set $\min(t)$ of minimal elements as defined in Sect. 1.2 is computable from the embedding π in Corollary 1.4.8. This gives a recursive procedure to compute the Foata normal form of t. The unicity of the normal form is also clear, since π is injective.

Corollary 1.4.10 *Free partially commutative monoids are cancellative.*

Proof: We already stated in the beginning that this is a consequence of Foata's theorem. Here, we see it directly from the embedding theorem since direct products of free monoids are cancellative for trivial reasons. $\square$

By the embedding theorem we may represent a trace as a tuple of words. This allows efficient computations on traces which are very easy to implement. In the next section we give some examples of such algorithms. The complexity of these algorithms depends, of course, of the number of components one uses to represent traces. This is the number of cliques which cover the dependence graph of the underlying dependence alphabet. In order to have a small number of cliques one usually tries to find a covering by maximal cliques. Note, however, that there are sometimes several different coverings by maximal cliques:

Example: Let (X, D) be the dependence alphabet which is given by Figure 1.8 and which represents a octahedron. Then there are two different coverings by maximal cliques. The first one corresponds to the embedding

$$\pi_1 : M(X, D) \hookrightarrow \{a, b, c\}^* \times \{a, d, e\}^* \times \{b, d, f\}^* \times \{c, e, f\}^*$$

The second one corresponds to the embedding

$$\pi_2 : M(X, D) \hookrightarrow \{d, e, f\}^* \times \{a, b, d\}^* \times \{a, c, e\}^* \times \{b, c, f\}^* \qquad \square$$

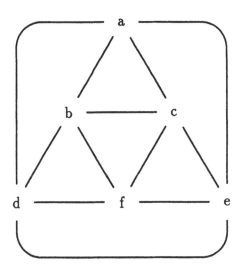

Figure 1.8. The octahedron with two different coverings by maximal cliques

1.5 Algorithms Based on the Embedding Theorem

The representation of a trace as a tuple of words leads to efficient algorithms which are suitable for implementation. We describe some of them in detail. We give algorithms which compute the Foata normal form, the lexicographic normal form, and we give a test whether one trace is a subtrace of another. This subtrace test is an anticipation of the last chapter where we investigate replacement systems over traces.

By (X, D) we mean a fixed, finite dependence alphabet. We choose a covering by cliques $(X, D) = (\bigcup_{i=1}^{k} X_i, \bigcup_{i=1}^{k} (X_i \times X_i))$ such that we obtain an embedding:

$$\pi : M(X, D) \hookrightarrow \prod_{i=1}^{k} X_i^* \quad , \quad X = \bigcup_{i=1}^{k} X_i \quad , \quad D = \bigcup_{i=1}^{k} (X_i \times X_i).$$

A letter of X is called in this section a *character* and a *string* means a sequence of characters. By a *step* we mean a subset of pairwise commuting letters of X. The Foata normal form will be computed as a *step sequence*. By a *tuple* we mean an *array* of k strings. An *index* means any number in $\{1, \ldots, k\}$. An *index set* denotes any subset of $\{1, \ldots, k\}$, and for each $a \in X$ we denote by component(a) the index set $\{i \in \{1, \ldots, k\} \mid a \in X_i\}$.

The term 1 is used to denote the empty string, the empty step-sequence, or the tuple of empty strings. We use some other notations and operations on strings, tuples etc. which meaning will become clear from the context. Every character is also viewed as a string, every string is also viewed as a tuple by using the embedding π. In fact this embedding may be realized by the following trivial algorithm:

Algorithm:
function π(s:string):tuple
var i:index;
var a:char;
var t:tuple:=1;
while s$\neq$1
do a:=first(s); s:=rest(s);
   for all i $\in$ component(a)
   do t[i]:=t[i]·a
   endfor
endwhile
return t
endfunction

The functions we shall use are always assumed to be strict in the sense that once an argument yields an undefined value, the whole function has an undefined value. This happens however only on incorrect inputs, i.e., if the input does not satisfy the specification of the function.

The next function computes the minimal elements of a trace, i.e., the first factor of the Foata normal form. For later use we consider firstly the case that some minimal letters are already known.

The algorithm looks somewhat complicated. Nevertheless the idea is simply that a letter $a \in X$ is minimal in a trace $t \in M(X, D)$ if and only if $a = \text{first}(t[i])$ for all $i \in$ component(a). The slightly complicated structure results only since we tried to be efficient.

Algorithm:
function min_1(m:step,t:tuple $\|$ t represents a trace with m $\subseteq$ min(t)):step;
var B:boolean:=true;
var a:char;
var F:step:=m;
var i,j:index;
var I,J:index-set;
I:={i $\in$\{1,...,k\}$|$ t[i] $\neq$ 1 and i $\notin$ component(a) for all a $\in$ m };
(" New minimal elements must have a component in I ")
while I $\neq \emptyset$

```
do choose any i ∈ I; I:=I \{i};
    a:=first(t[i]); J:=component(a) \{i};
    (" a is a candidat for a new minimal element ")
    while J ≠∅ and B
    do choose any j ∈ J;
        if j ∈ I and a=first(t[j])
        ("If t[j]=1 then first(t[j]) is not defined and t is not a trace. (Incorrect
        input!) The value of the function is undefined in this case since the
        functions are strict.")
        then I:=I\{j}; J:=J\{j}
        (" a is still a candidat ")
        else B:=false
        (" The candidat a is not minimal ")
        endif
    endwhile;
    if B
    then F:=F ∪{a}
    else B:=true
    endif
endwhile
return F
endfunction
```

If a tuple t does not represent a trace then the input of the function min_1 is not correct and it may yield the value undefined.

But since the input is not inspected at its whole length (for reasons of efficiency in the later algorithms) we can not force this. Moreover since there is some non-determinism in the algorithm above it may happen that on the same incorrect input the value of the function is sometimes undefined and sometimes a step. This results from the possible choices for $j \in J$ in the second while-loop. However on correct inputs as specified in the definition of min_1 everything works correctly.

The first factor of the Foata normal form is now:

Algorithm:
function min(t:tuple ‖ t represents a trace):step;
return min_1(∅,t) **endfunction**

If $l, t \in M(X, D)$ are traces and l is a prefix of t then we denote by $l^{-1}t$ the trace $r \in M(X, D)$ such that $t = lr$. We may view $l^{-1}t$ as a shorthand of a function in two arguments which is defined if l is a prefix of t and which value has the same type as t in this case.

The next algorithm computes the Foata normal form of a trace in linear time on a Turing machine. As a byproduct we obtain a test whether a tuple represents a trace since the value of that function will be defined if and only if the input satisfies the specification.

Algorithm:
function Foata-NF(t:tuple ‖ t represents a trace):step-sequence;
var F:step;
var S:step-sequence:=1;
while $t \neq 1$
do F:=min(t); S:=S·F; t:=$F^{-1}t$
endwhile
return S
endfunction

In order to compute the lexicographic normal form of a trace we assume that X is partially ordered such that every step is totally ordered. For example, X itself is totally ordered. The first letter of a step F with respect to this ordering is denoted by lexfirst(F).

Algorithm:
function lex-NF(t:tuple ‖ t represents a trace):string;
var a:char;
var s:string:=1;
var F:step:=min(t);
while $F \neq \emptyset$
do a:=lexfirst(F);
    s:=sa; t:=$a^{-1}t$;
    F:=min_1(F\{a},t)
endwhile
return s
endfunction

The algorithm above works again in linear time and it may also be used as a test whether a tuple represents a trace.

Our final goal in this section is to compute a factorization $t = \text{pre}(l)lu'$ for l being a subtrace of the trace t. More precisely, we define functions subtr(l, t)

and $\mathrm{pre}(l, t)$ of two tuple parameters l and t which represent traces. The value of $\mathrm{subtr}(l, t)$ is boolean and true if and only if l is a subtrace. The function $\mathrm{pre}(l, t)$ gives the value $\mathrm{pre}(l)$ at the first occurrence where we meet l as a subtrace of t. If l and t represent traces but l is no subtrace of t then the value of the function $\mathrm{pre}(l, t)$ will be undefined. But we do not say what happens if l or t is a tuple which does not represent a trace. (Such a specification is omitted since again we do not want to inspect the whole input if, for example, l is a prefix of t).

The idea of the function $\mathrm{pre}(l, t)$ (and $\mathrm{subtr}(l, t)$) is as follows: if l is a prefix of t then we set $\mathrm{pre}(l, t) = 1$. Otherwise we find an index $i \in \{1, \ldots, k\}$ such that $l[i]$ is no prefix of $t[i]$. If, in the latter case, we have $|l[i]| \geq |t[i]|$ then l is no subtrace of t. Otherwise, let a be the first letter of $t[i]$. Then we may compute recursively $\mathrm{pre}(l, t) = \mathrm{pre}(a, t) \, a \, \mathrm{pre}(l, (\mathrm{pre}(a, t)a)^{-1}t)$.

We first define a function which takes as input a letter $a \in X$, a trace $t \in M(X, D)$ with $a \in t$. It returns the value a if $a \in \min(t)$ and the value of some $b \in \min(t)$ such that $b \in \mathrm{pre}(a)$ otherwise. Note that this does not define a function in the pure sense, since we have nondeterministic output. The variable I below is used to ensure termination even on incorrect input.

Algorithm:

```
function minchar(a:char,t:tuple || t represents a trace such that a ∈ t):char;
var i:index;
var I:index-set:=∅;
var x:char:=a;
repeat choose any i ∈ component(x) \ I;
        x:=first(t[i]);
        I:=I∪{i})
until component(x)⊆I
return x
endfunction
```

The algorithm to compute minchar terminates in all cases since the index set I is augmented in each loop. But the value of minchar may be undefined on incorrect input. This happens if we want to assign $x := \mathrm{first}(t[i])$, but $t[i]$ is empty since, for example, $a \notin t$.

Let us sketch how to prove the correctness of the algorithm above. Let $n \geq 1$ be the number of repeat-loops and a_i (I_i respectively) the values of the variable x (variable I respectively) after the i-th round. Then we have $a_n \leq a_{n-1} \leq \ldots \leq a_1 \leq a_0 = a$. The crucial point is that if $a_{i-1} \neq a_i$ then $\mathrm{component}(a_i) \cap I_{i-1} = \emptyset$.

From the function above it is very easy to compute $\mathrm{pre}(a)$ for $a \in t$ where a is a letter and t is a trace. We denote this function by prechar, it is computed in time $O(|t|)$. Although we use the non-deterministic function minchar in it, it is clear that the result depends deterministically on the input.

Algorithm:
function prechar(a:char,t:tuple ‖ t represents a trace such that a $\in$ t):tuple;
var b:char:=minchar(a,t)
var s:tuple:=1;
while a $\neq$ b
do s:=s·b; t:=b^{-1}t
   b:=minchar(a,t)
endwhile
return s
endfunction

If we want to test whether a trace l is subtrace of a trace t, we may proceed as follows.

Algorithm:
function subtr(l,t:tuple ‖ l and t represent traces):boolean;
var i:index;
var a:char;
var B:boolean:=true;
while l is no prefix of t **and** B
do choose any i such that l[i] is not a prefix of t[i];
   if $|l[i]| \geq |t[i]|$
   then B:=false ("l is not a subtrace of t")
   else a:=first(t[i]);
      t:=(prechar(a,t)a)$^{-1}$t
   endif
endwhile
return B
endfunction

Now we handle the case where l is a subtrace of t. We can compute the subtrace of elements before the first occurrence of l as follows:

Algorithm:
function pre(l,t:tuple ‖ l $\subseteq$ t is a subtrace):tuple;
var i:index;

```
var a:char;
var s:tuple;
var p:tuple:=1;
while l is no prefix of t
do choose any i such that l[i] is not a prefix of t[i];
    a:=first(t[i]);
    s:=prechar(a,t)·a;
    p:=p·s;  t:=s⁻¹t
endwhile
return p
endfunction
```

If we know that l and t represent traces then the function $\mathrm{pre}(l,t)$ can be used as a subtrace test, too. Since then the value of the function is undefined if and only if l is no subtrace of t.

The time complexity of the function above is $O(|l \cdot t|)$. As in all our algorithms we viewed the alphabet X and the number k of covering cliques as fixed. If X and k are viewed as variable then we have to multiply the complexity by a factor $O(k \cdot \#X)$.

2. Recognizable and Rational Trace Languages

2.1 Recognizable and Rational Subsets of a Monoid

In this section we recall some very basic concepts of formal language theory in arbitrary monoids. We give no proofs. They are standard and can be found elsewhere, for example, in [Eil74].

Let M be any monoid. The set of rational expressions, RAT(M), is inductively defined as the smallest set satisfying the following conditions:

$$\text{If } E \subseteq M \text{ is finite} \qquad \text{then } E \ \in \ \text{RAT}(M).$$
$$\text{If } E_1, E_2 \in \text{RAT}(M) \quad \text{then } (E_1 \cup E_2) \ \in \ \text{RAT}(M),$$
$$E_1 \cdot E_2 \ \in \ \text{RAT}(M), \text{ and}$$
$$E_1^* \ \in \ \text{RAT}(M).$$

The symbols $\cup, \cdot, *$ are interpreted formally and brackets are used to avoid ambiguity. To each rational expression $E \in \text{RAT}(M)$ we associate a rational language $L(E) \subseteq M$ by giving the usual semantics to $\cup, \cdot, *$; i.e., $L(E) = E$ if $E \subseteq M$ is finite, $L(E_1 \cup E_2) = L(E_1) \cup L(E_2)$ (= union), $L(E_1 \cdot E_2) = L(E_1) \cdot L(E_2) = \{m_1 m_2 \in M \mid m_i \in L(M_i), i = 1, 2\}$ (= concatenation), and $L(E^*) = L(E)^*$ (= generated submonoid).

The set of rational languages of M is denoted by Rat(M).

Proposition 2.1.1 *Let $E \subseteq M$ be a subset of a monoid M. Then we have $E \in \text{Rat}(M)$ if and only if there is a finite alphabet X, a homomorphism $h : X^* \to M$, and a rational language $L \subseteq \text{Rat}(X^*)$ such that $h(L) = E$.* $\square$

Corollary 2.1.2 *The following assertions hold:*

i) $M \in \text{Rat}(M)$ if and only if M is finitely generated.

ii) If $f : M \to M'$ is a homomorphism of monoids and $E \in \text{Rat}(M)$ then $f(E) \in \text{Rat}(M')$.

iii) If $f : M \to M'$ is a surjective homomorphism then $E' \in \text{Rat}(M')$ if and only if $E' = f(E)$ for some $E \in \text{Rat}(M)$. $\square$

An M-*automaton* is a quadruple $A = (Z, \delta, q_0, F)$ where Z is the non-empty set of states, $\delta : Z \times M \to Z$, $\delta(q, m) = qm$ is the (partially defined) transition mapping such that $q1 = q$, $(qm)n = q(mn)$ for all $q \in Z$, $m, n \in M$, $q_0 \in Z$ is the initial state and $F \subseteq Z$ is the set of final states. An M-automaton is called *finite* if it has only finitely many states. The accepted language of an M-automaton $A = (Z, \delta, q_0, F)$ is the set $L(A) = \{m \in M \mid q_0 m \in F\}$. If $L = L(A)$ for some finite automaton then L is called *recognizable*. The set of recognizable subsets of M is denoted by $\text{Rec}(M)$.

Let $L \subseteq M$ be any subset then the following automaton $A_L = (Z, \delta, q_0, F)$ accepts L where $Z = \{L(x) \mid x \in M\}$, $L(x) = \{y \in M \mid xy \in L\}$ for $x \in M$, $\delta(L(x), y) := L(xy)$ for $x, y \in M$, $q_0 = L(1)$ and $F = \{L(x) \mid x \in L\}$. The automaton A_L is called the *minimal automaton* of L.

Let $y, y' \in M$ then we have $\delta(L(x), y) = \delta(L(x), y')$ for all $x \in M$ if and only if for all $x, z \in M$ the assertions $xyz \in L$ and $xy'z \in L$ are equivalent. We write $y \equiv_L y'$ in this case. The relation $\equiv_L$ is a congruence and the quotient monoid $\text{Synt}(L) = M/_{\equiv_L}$ is called the *syntactic monoid* of M. Note that for the canonical morphism $h : M \to M/_{\equiv_L}$ we have $L = h^{-1}h(L)$.

Proposition 2.1.3 *Let $L \subseteq M$ be a subset of a monoid M. Then the following assertions are equivalent:*

i) *The subset $L \subseteq M$ is recognizable.*

ii) *The minimal automaton A_L is finite.*

iii) *The syntactic monoid $\text{Synt}(L)$ is finite.* $\square$

Corollary 2.1.4 *Let $f : M \to M'$ be a homomorphism of monoids. If $L' \in \text{Rec}(M')$ then $f^{-1}(L') \in \text{Rec}(M)$. If f is surjective then $f^{-1}(L') \in \text{Rec}(M)$ implies $L' \in \text{Rec}(M')$, too.* $\square$

Corollary 2.1.5 *The recognizable subsets, $\text{Rec}(M)$, form a boolean algebra: they are closed under finite union, finite intersection and complementation.* $\square$

If $M = X^*$ is a finitely generated monoid then Kleene's theorem asserts $\text{Rec}(X^*) = \text{Rat}(X^*)$. In general, both sets are incomparable. However, in finitely generated monoids recognizable subsets are rational:

Corollary 2.1.6 *A monoid M is finitely generated if and only if*

$$\text{Rec}(M) \subseteq \text{Rat}(M). \qquad \square$$

Probably the most simple examples where one has a proper containment of $\text{Rec}(M)$ in $\text{Rat}(M)$ are given by infinite groups. Indeed, let $M = G$ be any group and $U \subseteq G$ be a subgroup. Then U is recognizable in G if and only if it is of finite index in G. Hence, the rational set $\{1\} \subseteq G$ is recognizable if and only if G is a finite group. This observation leads to an elegant way to deduce that the rational set $(1,1)^* \subseteq \mathbb{N} \times \mathbb{N}$ is not recognizable in $\mathbb{N} \times \mathbb{N}$. In particular, we need no pumping lemma to see this: Let $f : \mathbb{N} \times \mathbb{N} \to \mathbb{Z}$ be the surjective homomorphism onto the group of integers defined by $f(1,0) = 1$, $f(0,1) = -1$; then $f^{-1}(0) = (1,1)^*$. Thus $(1,1)^*$ would be recognizable only if $\{0\} \in \text{Rec}(\mathbb{Z})$: But $\mathbb{Z}$ is an infinite group.

2.2 Closure Properties of Recognizable Trace Languages

Rational trace languages are called *existentially regular* in [AR86] or [AW86] and regular in [Sak87]. On the other hand, regular for trace languages means recognizable in [Och85] or [Zie87] and this is called *consistently regular* in [AR86], [AW86]. Thus, regular has different meanings. Although the meaning of consistently and existentially regular trace languages is also clear, we prefer to continue to speak of rational and recognizable languages. The notation of regular sets is reserved for rational (or recognizable) subsets of finitely generated free monoids. This seems also to be in accordance with the classical terminology and avoids any confusion.

Let $M = M(X, D)$ be the trace monoid of a fixed finite dependence alphabet (X, D). As before, we write $(u, v) \in I$ for $u, v \in M$ if $\text{alph}(u) \times \text{alph}(v) \subseteq I = X \times X \setminus D$.

Clearly, finite trace languages are recognizable and $\text{Rec}(M)$ is closed under union. It belongs also to folklore that if M is not free then $\text{Rec}(M)$ is not closed under the star operator and hence $\text{Rec}(M)$ is properly contained in $\text{Rat}(M)$. This can be derived, for example, from the fact that $(1,1)^* \notin \text{Rec}(\mathbb{N} \times \mathbb{N})$.

The first non-trivial result in this area was obtained by M. Fliess. It states that $\text{Rec}(M)$ is closed under concatenation:

Proposition 2.2.1 ([Fli74, Prop. 2.2.15]) *Let $A, B \in \text{Rec}(M)$ be recognizable trace languages then $AB \in \text{Rec}(M)$.*

The original proof by Fliess using Hankel matrices is quite involved. The proof we give here is a slight modification of the one given in [CP85]. This modification is used to develop a linear time algorithm which computes a factorization $z = xy$ with $x \in A$, $y \in B$ for given trace $z \in AB$, c.f. Corollary 2.2.3 below, see also [AW86]. The idea behind the proof is to construct first (implicitly) a non-deterministic automaton which recognizes AB and then to define a real M-automaton by the usual power set construction.

Proof: Recall that the states of the minimal automaton which recognizes A (B resp.) are denoted by $A(x)$ ($B(x)$ resp.) with $x \in M$. Consider the following finite set: $Q = \{(A(x), B(y), \text{alph}(y)) \mid x, y \in M\}$.

For a trace $z \in M$ let $V(z) = \{(A(x), B(y), \text{alph}(y)) \mid z = xy\}$ and $Z = \{V(z) \mid z \in M\}$. The set Z is a subset of the finite powerset of Q. There is a natural way to define an operation δ of M on Z, namely by $\delta(V(z), m) = V(zm)$. However, the problem is to show that this is well-defined; i.e., if $V(z_1) = V(z_2)$ for traces $z_1, z_2 \in M$ then we have to show $V(z_1 m) = V(z_2 m)$ for all $m \in M$. In fact, it is enough to show $V(z_1 a) = V(z_2 a)$ for all $a \in X$.

Assume for a moment that we have shown this already then we are through. Indeed,

$$A = (Z, \delta, V(1), \{V(z) \mid z \in AB\})$$

is a finite M-automaton which recognizes AB. (Note that $V(z_1) = V(z_2)$ with $z_2 \in AB$ implies $z_1 \in AB$ by definition of $V(z)$ for $z \in M$.)

In order to prove that the operation $\delta : Z \times M \to Z$ is well-defined we are going back to work on the set Q.

Define a mapping

$$\eta : Q \times X \to Q \text{ by } \eta((A(x), B(y), \text{alph}(y)), a) = (A(x), B(ya), \text{alph}(ya))$$

and a partial mapping

$$\epsilon : Q \times X \to Q$$

by $\epsilon((A(x), B(y), \text{alph}(y)), a) = (A(xa), B(y), \text{alph}(y))$ if y and a are independent (i.e. $(y, a) \in I$)
and undefined otherwise.

(One could view the relation $\eta \cup \epsilon \subseteq Q \times X \times Q$ as edges of a non-deterministic automaton. But we do not insist on this explicitly.)

The remaining claim that $V(z_1) = V(z_2)$ implies $V(z_1 a) = V(z_2 a)$ for all $z_1, z_2 \in M$, $a \in X$, is a consequence of the following lemma.

Lemma 2.2.2 *For all $z \in M$ and $a \in X$ we have*

$$V(za) = \eta(V(z), a) \cup \epsilon(V(z), a).$$

Proof: From the definition of η and ϵ we have $\eta(V(z), a) \cup \epsilon(V(z), a) \subseteq V(za)$. Let $(A(x), B(y), \text{alph}(y)) \in V(za)$. Then $xy = za$ and by Levi's Lemma we have $x = ru$, $y = vs$, $z = rv$, $a = us$ for some $r, u, v, s \in M$ with $(u, v) \in I$. It follows $(A(r), B(v), \text{alph}(v)) \in V(z)$. There are two cases, first: $u = 1$ and $s = a$, second: $u = a$ and $s = 1$. In the first case $(A(x), B(y), \text{alph}(y)) \in \eta(V(z), a)$ in the second $(A(x), B(y), \text{alph}(y)) \in \epsilon(V(z), a)$. $\square$

Corollary 2.2.3 *Let $A, B \in \text{Rec}(M)$ be recognizable trace languages. Then there is a linear time algorithm which decides on input of a trace $z \in M$ whether $z \in AB$ and if $z \in AB$ then it computes $x \in A$, $y \in B$ such that $z = xy$.*

Proof: Let $z \in M$ be a trace. We may assume that z is given in the form $z = a_1 \ldots a_m$ with $m \geq 0$ and $a_i \in X$ for $1 \leq i \leq m$. In time $O(m)$ we compute the sequence $V(1), V(a_1), V(a_1 a_2), \ldots, V(a_1 \ldots a_m)$, and we have $z \in AB$ if and only if $V(z) = V(a_1 \ldots a_m) \in F$. Assume $z \in AB$ then choose $q_m = (A(x'), B(y'), \text{alph}(y')) \in V(z)$ such that $x' \in A$, $y' \in B$. By Lemma 2.2.2 there is a sequence $(q_m, \ldots, q_1, q_0)$ such that $q_0 = (A(1), B(1), \emptyset)$ and $q_i = \eta(q_{i-1}, a_i)$ or $q_i = \epsilon(q_{i-1}, a_i)$ for all $1 \leq i \leq m$. Again, this sequence can be computed in $O(m)$ steps, starting with q_m. Finally, set $x_0 = y_0 = 1$ and for $i = 1, \ldots, m$ define traces x_i, y_i inductively as follows: if $q_i = \eta(q_{i-1}, a_i)$ then $x_i := x_{i-1}, y_i := y_{i-1} a_i$ else $x_i := x_{i-1} a_i, y_i := y_{i-1}$. Thus, the whole procedure to compute x_m, y_m is done in time $O(m)$. It suffices to show $z = x_m y_m$ and $x_m \in A$, $y_m \in B$. But this is clear since for all $0 \leq i \leq m$ we have $a_1 \ldots a_i = x_i y_i$ and $q_i = (A(x_i), B(y_i), \text{alph}(y_i))$. Hence $z = x_m y_m$, and $(A(x_m), B(y_m), \text{alph}(y_m)) = (A(x'), B(y'), \text{alph}(y'))$ and it follows $x_m \in A$, $y_m \in B$. $\square$

A trace $t \in M$ is called *connected* if the dependence graph of t is connected or, what amounts to the same, if the letters of $\text{alph}(t)$ induce a connected subgraph in the graph of the dependence alphabet (X, D). Every trace $t \in M$ has a partition into connected components $t = t_1 \dot\cup \ldots \dot\cup t_n$ where $t_i \subseteq t$ are non-empty connected subtraces and $(t_i, t_j) \in I$ for $1 \leq i, j \leq n$, $i \neq j$. Obviously, every trace is the product of its connected components. Let $L \subseteq M$ be any trace language then we define the language of its connected components by

$$\text{Con}(L) = \{x \in M \mid x \text{ is a connected component of some } t \in L\}.$$

Note that we never have $1 \in \text{Con}(L)$ although $1 \in M$ is a connected trace.

Proposition 2.2.4 ([Och85, Lemma 8.1]) *Let $L \in \text{Rec}(M)$ be a recognizable trace language then we have $\text{Con}(L) \in \text{Rec}(M)$.*

Proof: Let $t \in M$ be any trace. Then $y \in M$ is a connected component of t if and only if $\text{alph}(y)$ is a non-empty connected subgraph in (X, D) and $t = yz$ for some $z \in M$ with $\text{alph}(y) \times \text{alph}(z) \in I$. Hence $\text{Con}(L)$ is a finite union of languages of type $L_{Y,Z} = \{y \in M \mid \text{alph}(y) = Y \text{ and } \exists z \in M : yz \in L, \text{alph}(z) \subseteq Z\}$. Since the set $\{y \in M \mid \text{alph}(y) = Y\}$ is recognizable, it is enough to show that $L_Z = \{y \in M \mid \exists z \in M : yz \in L, \text{alph}(z) \subseteq Z\}$ is recognizable for every subalphabet $Z \subseteq X$.

But this is easy: Let $A = (Q, \delta, q_0, F)$ be any finite M-automaton with $L(A) = L$. Changing the set of final states we obtain effectively a finite M-automaton $A_Z = (Q, \delta, q_0, \{q \in Q \mid \exists z \in M : \delta(q, z) \in F \text{ and } \text{alph}(z) \subseteq Z\})$ which recognizes L_Z. $\square$

The next lemma is due to R. Cori and D. Perrin. It is a generalization of the Levi Lemma for traces.

Lemma 2.2.5 ([CP85, Cor. 1.4]) *Let* $x, y, z_1, \ldots, z_m \in M$ *be traces,* $m \geq 1$ *such that* $xy = z_1 \ldots z_m$. *Then there are* $r_i, s_i \in M$, $1 \leq i \leq m$ *such that* $x = r_1 \ldots r_m$, $y = s_1 \ldots s_m$, $z_i = r_i s_i$ *for* $1 \leq i \leq m$ *and* $(r_j, s_i) \in I$ *for* $1 \leq i < j \leq m$.

Proof: Let $t = xy = z_1 \ldots z_m$. The proof results directly from Figure 2.1.

Formally, view $x, y, z_1, \ldots, z_m \subseteq t$ as subtraces of t. Define $r_i = x \cap z_i$, $s_i = y \cap z_i$ for $1 \leq i \leq m$. It follows $x = r_1 \ldots r_m$, $y = s_1 \ldots s_m$, and $z_i = r_i s_i$ for $1 \leq i \leq m$. Let $1 \leq i < j \leq m$, then $r_j \cap \text{pre}(s_i) = \emptyset$ since $t = (r_1 s_1) \ldots (r_i s_i) \ldots (r_j s_j) \ldots (r_m s_m)$. Analogously, since $t = r_1 \ldots r_m s_1 \ldots s_m$ we have $r_j \cap \text{suf}(s_i) = \emptyset$. Hence $r_j \subseteq \text{ind}(s_i)$, and this implies $(r_j, s_i) \in I$. $\square$

The following proposition has been shown independently by E. Ochmanski and Y. Métivier. The paper of Métivier is based on previous results obtained by Cori/Perrin, [CP85, Thm. 2.1], and Cori/Métivier, [CM85, Thm. 2.7].

Proposition 2.2.6 ([Och85, Lemma 8.2], [Mét86, Thm. 2.3])
Let $L \in \text{Rec}(M)$ *be a recognizable trace language which consists of connected traces only, i.e.,* $L \setminus \{1\} = \text{Con}(L)$. *Then we have* $L^* \in \text{Rec}(M)$.

Proof[Mét86]: For a trace $x \in M$ let $V(x)$ be set of tuples

$$(L(r_1), \ldots, L(r_n), \text{alph}(r_1), \ldots, \text{alph}(r_n), \text{alph}(p_0), \ldots, \text{alph}(p_n))$$

where $n \geq 0, r_i, p_i \in M$, $r_i \neq 1$, $p_i \in L^*$ for $0 \leq i \leq n$, $\text{alph}(r_i) \neq \text{alph}(r_j)$ for $1 \leq i \neq j \leq n$, and $x = p_0 r_1 p_1 \ldots r_n p_n$.

Since $\text{alph}(r_i) \neq \text{alph}(r_j)$ for $i \neq j$, the number $n \geq 0$ is bounded by some constant independent of x; and hence $\{V(x) \mid x \in M\}$ is a finite set of finite sets. Therefore it is enough to show that $V(x) = V(x')$ implies $L^*(x) = L^*(x')$ for all $x, x' \in M$. Let $x, x', y \in M$ such that $V(x) = V(x')$ and $y \in L^*(x)$. By symmetry it is enough to prove $y \in L^*(x')$.

Since $y \in L^*(x)$, we have $xy = z_1 \ldots z_m$ for some $m \geq 0$, $z_i \in L$, $1 \leq i \leq m$. By Lemma 2.2.5 there are $r_i, s_i \in M$, $1 \leq i \leq m$ such that $x = r_1 \ldots r_m$, $y = s_1 \ldots s_m$, $z_i = r_i s_i$ for $1 \leq i \leq m$ and $(r_j, s_i) \in I$ for $1 \leq i < j \leq m$.

Now, for $i < j$ define blocks $p_{ij} = r_i \ldots r_j$ if $s_i \ldots s_j = 1$ and $q_{ij} = s_i \ldots s_j$ if $r_i \ldots r_j = 1$. Note $p_{ij}, q_{ij} \in L^*$ for all $i < j$ where these traces are defined.

By a conversion of indices, we obtain for some $n \geq 0$:

$$
\begin{aligned}
x &= p_0 r_1 p_1 \ldots r_n p_n, & & p_0 \in L^*, r_i \neq 1, p_i \in L^*, \\
y &= q_0 s_1 q_1 \ldots s_n q_n, & & q_0 \in L^*, s_i \neq 1, q_i \in L^*, \\
r_i, s_i &\in L, \text{ and} \\
(r_i p_i, q_0 s_1 q_1 \ldots s_{i-1} q_{i-1}) &\in I & & \text{for all } 1 \leq i \leq n
\end{aligned}
$$

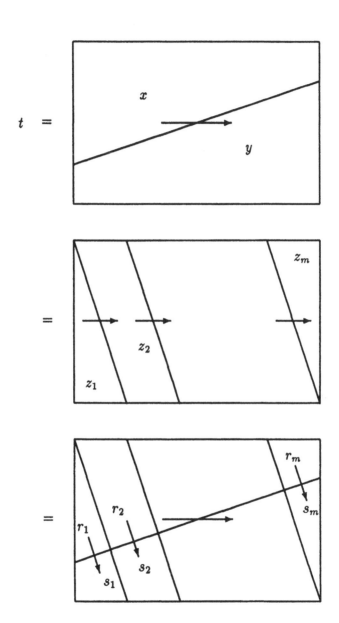

Figure 2.1. The generalized Levi Lemma for traces

Assume $\mathrm{alph}(r_i) = \mathrm{alph}(r_j)$ for some $i \leq j$. Since $r_i s_i \in L$ is a connected trace and this depends on $\mathrm{alph}(r_i s_i)$ only, the trace $r_j s_i$ is connected, too. Since $(r_j, s_i) \in I$ for $i < j$ and $r_j \neq 1$, $s_i \neq 1$, we must have $i = j$.

Thus, $(L(r_1), \ldots L(r_n), \mathrm{alph}(r_1), \ldots, \mathrm{alph}(r_n), p_0, \ldots, p_n) \in V(x)$. Since $V(x) = V(x')$ we find $r_1' \ldots r_n' \in M \setminus \{1\}$, $p_0', \ldots, p_n' \in L^*$ such that $x' = p_0' r_1' p_1' \ldots r_n' p_n'$ and $L(r_i) = L(r_i')$, $\mathrm{alph}(r_i) = \mathrm{alph}(r_i')$, $\mathrm{alph}(p_i) = \mathrm{alph}(p_i')$ for all $1 \leq i \leq n$.
Consider:

$$x'y = p_0' r_1' p_1' \ldots r_n' p_n' q_0 s_1 q_1 \ldots s_n q_n = p_0' q_0 (r_1' s_1) p_1' q_1 \ldots (r_n' s_n) p_n' q_n \in L^*$$

This shows $y \in L^*(x')$ and hence the result. $\square$

Corollary 2.2.7 *Let $L \in \mathrm{Rec}(M)$ be any recognizable trace language. If $\mathrm{Con}(L) \subseteq L^*$ then we have $L^* \in \mathrm{Rec}(M)$.*

Proof: If $\mathrm{Con}(L) \subseteq L^*$ then it follows $((\mathrm{Con}(L))^* \subseteq L^*$. Since the opposite inclusion is true for any language, we have $L^* = ((\mathrm{Con}(L))^*$ which is recognizable by the previous proposition. $\square$

Remark 2.2.8 i): The condition $\mathrm{Con}(L) \subseteq L^*$ is satisfied if, for example, L is prefix closed, since every connected component of a trace $t \in M$ is a prefix of t.

ii): If L is recognizable and $\mathrm{Con}(L) \subseteq L^*$ then L^* is a submonoid which is generated by a recognizable set of connected traces. By Proposition 2.2.6 this is enough to ensure that L^* is recognizable. On the other hand this is not a necessary condition: Consider $L = \{(m, n) \in \mathbb{N} \times \mathbb{N} \mid m + n \geq 2\}$ then $L \cup \{(0,0)\} = L^*$ is recognizable but not generated by connected elements of $\mathbb{N} \times \mathbb{N}$. In fact, up to now no convenient characterization of recognizable submonoids is known. Here we content ourselves to state the following:

Corollary 2.2.9 ([CM85, Thm. 2.7],[Dub86, Thm. 4.3.2])
Let $t \in M$ be a trace. Then we have $t^* \in \mathrm{Rec}(M)$ if and only if t is connected.

Proof: If t is connected then $t^* \in \mathrm{Rec}(M)$ by 2.2.6. If t is not connected then we may write $t = uv = vu$ for some $u, v \in M$, $u \neq 1$, $v \neq 1$ with $(u, v) \in I$. Consider the homomorphism $f : \mathbb{N} \times \mathbb{N} \to M$, $(1,0) \mapsto u$, $(0,1) \mapsto v$. Then $f^{-1}(t^*) = (1,1)^* \subseteq \mathbb{N} \times \mathbb{N}$ and $(1,1)^*$ is not recognizable in $\mathbb{N} \times \mathbb{N}$. $\square$

2.3 Ochmanski Theory

This section contains the result of E. Ochmanski, [Och85], which characterizes recognizable trace languages in terms of so-called recognizable expressions. Formally, the classical Kleene Theorem is a corollary of Ochmanski's theorem. Nevertheless, (the non-trivial direction) of Kleene's theorem is crucial below.

Following Ochmanski, we introduce the *concurrent iteration* by $L^{co} = (\mathrm{Con}(L))^*$ for $L \subseteq M$, i.e., L^{co} is the submonoid generated by all connected components of traces in L. Analogously to rational expressions the set of recognizable expressions, $\mathrm{REC}(M)$, is defined. We only replace the (formal symbol) star, $*$, by the symbol of concurrent iteration, co. If $E \in \mathrm{REC}(M)$ is a recognizable expression then $L(E) \subseteq M$ means the associated language. In particular, $L(E^{co}) := L(E)^{co} = (\mathrm{Con}(L(E)))^*$. The preceeding section may be summarized by saying $L(E) \subseteq \mathrm{Rec}(M)$ for all $E \in \mathrm{REC}(M)$. Ochmanski's theorem says that the converse is also true.

Theorem 2.3.1 ([Och85, Thm. 9]) *Let $L \subseteq M$ be a trace language. $M = M(X, D)$ for some finite dependence alphabet (X, D). Then L is recognizable, $L \in \mathrm{Rec}(M)$, if and only if there is a recognizable expression $E \in \mathrm{REC}(M)$ such that $L = L(E)$.*

The proof of 2.3.1 follows from the next two lemmas which concern lexicographic normal forms. Let $<$ be any linear ordering of the alphabet X. Thus, we may assume $X = \{a_1, \ldots, a_m\}$ and $a_1 < \cdots < a_m$. For each trace $t \in M$ we denote by $\hat{t} \in X^*$ the lexicographic normal form of t, i.e., the minimal word with respect to the lexicographic order which represents the trace t. Then $\hat{M} = \{\hat{t} \in X^* \mid t \in M\}$ is in canonical bijection with M. By the proposition of Anisimov and Knuth, see Proposition 1.2.2, $\hat{M}$ is exactly the rational set of words which contain no factor of the form bua with $a < b$ and $(a, bu) \in I$.

Lemma 2.3.2 ([Och85, Lemma 4.2]) *Let $t \in M$ be a trace such that the square of its lexicographic normal form is in lexicographic normal form, i.e., $\hat{t}^2 \in \hat{M}$. Then t is connected.*

Proof: Assume t would not be connected. Write $t = uv$ with $u, v \in M \setminus \{1\}$, $(u, v) \in I$. Let $U = \mathrm{alph}(u)$ and $V = \mathrm{alph}(v)$. Then $\hat{t} \in X^*$ starts with a letter in U or V, say U, and $\hat{t}$ has the following form $\hat{t} = b_1 w_1 b_2 w_2 \ldots b_n w_n$ with $n \geq 2$, $b_i \in U \cup V$ and $b_i w_i \in V^*$ for i even and $b_i w_i \in U^*$ for i odd, $1 \leq i \leq n$.

Since $\hat{t}$ is in lexicographic normal form we have $b_1 < b_2 < \cdots < b_n$. If n is even then $(\hat{t})^2$ contains the factor $b_n w_n b_1$ with $b_1 < b_n$ and $(b_n w_n, b_1) \in I$. But this is impossible. If n is odd then $(\hat{t})^2$ contains the factor $b_n w_n b_1 w_1 b_2$ with

$b_2 < b_n$ and $(b_n w_n b_1 w_1, b_2) \in I$. Again this is impossible. Hence $(\hat{t})^2 \notin \hat{M}$ which is a contradiction. $\square$

Let us now define a mapping $\pi : \mathrm{RAT}(X^*) \to \mathrm{REC}(M)$ on expressions inductively as follows:

If $E \subseteq X^*$ is finite then $\pi(E)$ means the finite image in M. If $E, E_1, E_2 \in \mathrm{RAT}(X^*)$ and $\pi(E), \pi(E_1), \pi(E_2)$ are already defined then set $\pi(E_1 \cup E_2) = \pi(E_1) \cup \pi(E_2)$, $\pi(E_1 \cdot E_2) = \pi(E_1) \cdot \pi(E_2)$ and $\pi(E^*) = (\pi(E))^{\infty}$.

Let $p : X^* \to M$ be the canonical projection and $E \in \mathrm{RAT}(X^*)$ then we have $pL(E) \subseteq L\pi(E)$, but no equality in general. The point is that there is equality on lexicographic normal forms:

Lemma 2.3.3 *Let $E \in \mathrm{RAT}(X^*)$ be a rational expression such that $L(E) \subseteq \hat{M}$. Then we have $pL(E) = L\pi(E)$.*

Proof: By structural induction: The formula is obvious for finite E. If $E = E_1 \cup E_2$ and $L(E) \subseteq \hat{M}$ then $L(E_1), L(E_2) \subseteq \hat{M}$. Hence by induction, $pL(E) = pL(E_1 \cup E_2) = pL(E_1) \cup pL(E_2) = L\pi(E_1) \cup L\pi(E_2) = L\pi(E)$. Analogously for $E = E_1 E_2$. Finally let $E = E_1^*$ for some $E_1 \in \mathrm{RAT}(X^*)$. Again, if $L(E_1^*) \subseteq \hat{M}$ then $L(E_1) \subseteq \hat{M}$ and $L(E_1)^* \subseteq \hat{M}$. Hence we may use induction and Ochmanski's Lemma tells us that $pL(E_1)$ consists of connected traces only. Therefore:

$$pL(E_1^*) = (pL(E_1))^* = (\mathrm{Con}(pL(E_1)))^* = (pL(E_1))^{\infty} =$$

$$(L\pi(E_1))^{\infty} = L((\pi(E_1)^{\infty}) = L\pi(E_1^*). \quad \square$$

Proof of 2.3.1: If $E \in \mathrm{REC}(M)$ then $L(E) \in \mathrm{Rec}(M)$ by Sect. 2.2. Let $L \in \mathrm{Rec}(M)$ be any recognizable language. Then the inverse image $p^{-1}(L)$ is recognizable in X^*. Since $\hat{M} \subseteq X^*$ is recognizable, too, $p^{-1}(L) \cap \hat{M}$ is a recognizable subset of X^*; and by Kleene's Theorem $p^{-1}(L) \cap \hat{M} = L(E)$ for some rational expression $E \in \mathrm{RAT}(X^*)$. Since $L(E) \subseteq \hat{M}$, we have $L(\pi(E)) = pL(E) = p(p^{-1}(L) \cap \hat{M}) = L$ by Lemma 2.3.3. Since $\pi(E) \in \mathrm{REC}(M)$, the Theorem follows. $\square$

Corollary 2.3.4 *The set of recognizable trace languages of M is in canonical one-to-one correspondence with the set of regular subsets of $\hat{M}$. $\square$*

2.4 Zielonka Theory

The automata we considered so far have the disadvantage that independent actions can not really act concurrently on the state set. This led W. Zielonka [Zie87] to introduce asynchronous automata which have the ability of simultaneous execution of independent actions. The main result of [Zie87] states that a subset of a trace monoid is recognizable if and only if it is accepted by some finite asynchronous automaton. Our presentation of Zielonka's result is based on the original paper [Zie87] and on a technical report of Cori/Métevier [CM87]. Almost all results of this section can be found there. We do not claim any originality. We also consider *asynchronous cellular automata*. These automata were introduced in [Zie89] and also (independently) used in [Die90a]. Although closely related to asynchronous automata, asynchronous cellular automata seem to have a slightly simpler construction. These automata will be used in the last chapter for the design of efficient algorithms to solve word problems over traces.

Let (X, D) be a finite dependence alphabet with independence relation $I = X \times X \setminus D$ and $M = M(X, D)$. A (finite) *asynchronous automaton* over M is a (finite) M-automaton $A = (Z, \delta, q_0, F)$ which satisfies the following additional conditions:

1) The state set Z is a cartesian product $Z = \prod_{i=1}^{m} Z_i$.

2) With each $a \in X$ there is a associated an index set $V(a) \subseteq \{1, \ldots, m\}$ such that $V(a) \cap V(b) = \emptyset$ if and only if $(a, b) \in I$.

3) The partially defined transition mapping $\delta : Z \times M \to M$ is given by a collection of partially defined mappings

$$\{\delta_a : \prod_{i \in V(a)} Z_i \to \prod_{i \in V(a)} Z_i \mid a \in X\}.$$

This means, for a state $(z_1, \ldots, z_m) \in Z$ and a letter $a \in X$ the next-state $\delta((z_1, \ldots, z_m), a)$ is defined if and only if $\delta_a((z_i)_{i \in V(a)})$ is defined and in this case the j-th component of $\delta((z_1, \ldots, z_n), a)$ is $\delta((z_1, \ldots, z_m), a)_j = (\delta_a((z_i)_{i \in V(a)}))_j$ for $j \in V(a)$ and $\delta((z_1, \ldots, z_m), a)_j = z_j$ otherwise, $1 \leq j \leq m$. Thus, the action of a changes the components in $V(a)$ only.

Remark 2.4.1 Every asynchronous automaton can be transformed into an asynchronous automaton where the transition mapping is defined everywhere. This amounts to add to each component Z_i, $1 \leq i \leq m$, a new point, say $\{\infty\}$. But observe that for large $m > 1$ the state set $\prod_{i=1}^{m} (Z_i \dot\cup \{\infty\})$ has considerably more points than $\prod_{i=1}^{m} Z_i$. This is the reason to prefer partial transition mappings. $\square$

The aim of this section is to prove the following deep theorem of Zielonka:

Theorem 2.4.2 ([Zie87]) *Let $L \subseteq M$ be a (recognizable) trace language then there exists a (finite) asynchronous automaton which accepts L.*

In order to prove this result we use another notion of M-automata. An M-automaton $U = (Z, \delta, q_0, F)$ is called *asynchronous cellular* if the following two condition are satisfied:

1) The state set Z is a cartesian product $Z = \prod_{x \in X} Z_x$

2) The partially defined transition mapping δ is given by a collection of partially defined mappings $\{\delta_a : (\prod_{b \in D(a)} Z_b) \to Z_a \mid a \in X\}$ where $D(a) = \{b \in X \mid (a, b) \in D\}$.

This means for $(z_x)_{x \in X}$ and $a \in X$ the next state $\delta_a((z_x)_{x \in X}, a)$ is defined if and only if $\delta_a((z_b)_{b \in D(a)})$ is defined and in this case we have $\delta((z_x)_{x \in X}, a)_a = \delta_a((z_b)_{b \in D(a)})$ and $\delta((z_x)_{x \in X}, a)_y = z_y$ for $y \neq a$. Thus, the action of a changes only the a-component of $(z_x)_{x \in X}$, but the new value of z_a depends on the b-components for $b \in D(a)$.

Note that the classes of asynchronous automata and asynchronous cellular automata are incomparable. (It is of course also possible to give a common generalization of both types of automata. This is not needed here and left to the reader.) However, since under an action $a \in X$ changes only the a-component of the state set and $b \notin D(a)$ for $(a, b) \in I$, asynchronous cellular automata also have the ability of simultanously execution of independent actions.

This is a typical situation where we allow common read but exclusive write. Moreover there is a simple standard construction which transforms an asynchronous cellular automaton into an asynchronous one by blowing up the state set. This is the first proposition of this section.

Proposition 2.4.3 *Let $(X, D) = \bigcup_{i=1}^{m} (X_i, X_i \times X_i)$ be a covering by cliques and let $U = (\prod_{a \in X} Z_a, \delta, q_0, F)$ be an asynchronous cellular automaton. Then there is an asynchronous automaton which accepts the same trace language as U and which has state set $\prod_{i=1}^{m} (\prod_{a \in X_i} Z_a)$. In particular, the asynchronous automaton is finite if U is finite.*

Proof: Let $Y = \{(a, i) \mid 1 \leq i \leq m, a \in X_i\}$. Consider the diagonal embedding

$$d : \prod_{a \in X} Z_a \to \prod_{i=1}^{m} (\prod_{a \in X_i} Z_a) = \prod_{y \in Y} Z_y, \quad (z_a)_{a \in X} \to (z_a)_{y \in Y} \text{ for } y = (a, i)$$

Viewing d as an inclusion the partial transition mapping $\delta : (\prod_{a \in X} Z_a) \times M \to$
$\prod_{a \in X} Z_a$ can also be viewed as a partial mapping $\delta : (\prod_{y \in Y} Z_y) \times M \to \prod_{y \in Y} Z_y$. Thus,
$A = (\prod_{y \in Y} Z_y, \delta, d(q_0), d(F))$ as an M-automaton which, by the very definition,
accepts the same trace language as U. The automaton A looks like U, but from
each original component of the state set of U there are now several copies in A.
Defining $Z_i = \prod_{a \in X_i} Z_a$ and $V(a) = \{i \in \{1, \ldots, m\} \mid a \in X_i\}$ for $a \in X$, we see
that A has the structure of an asynchronous automata. This follows since we
have $\prod_{y \in Y} Z_y = \prod_{i=1}^{m} Z_i$ and $V(a) \cap V(b) = \emptyset$ if and only if $(a, b) \in I$. $\Box$

Let $t \in M$ be a trace and $p_1, p_2 \subseteq t$ be prefixes of t, i.e., $t = p_1 s_1 = p_2 s_2$ for
some $s_1, s_2 \in M$. Then $p_1 \cup p_2 \subseteq t$ is a prefix of t. If $\max(p_1) \subseteq A$, $\max(p_2) \subseteq A$
for some $A \subseteq X$ then we have $\max(p_1 \cup p_2) \subseteq A$, too. This observation allows
to define the maximal prefix of a trace t such that the set of maximal letters of
the prefix lies in a given subset $A \subseteq X$. In the following we denote this prefix by
$\partial_A(t)$. If $A = \{a\}$ then we simply write $\partial_a(t)$. Note that this prefix is non-empty
if and only if $a \in \text{alph}(t)$. If $\partial_a(t) \neq 1$ then for the maximal point of t with label
$a \in X$ we can also write in our previous notation $\partial_a(t) = \text{pre}(a) \cdot a$.

It is easy to see that $\partial_A(t) = \bigcup_{a \in A} \partial_a(t)$ and hence $\partial_{A \cup B}(t) = \partial_A(t) \cup \partial_B(t)$. For
a letter $a \in X$ let $D(a) = \{b \in X \mid (a, b) \in D\}$, then we have $\partial_a(ta) = \partial_{D(a)}(t)a$.
These simple identities will be used frequently in the following.

Definition: A mapping $\varphi : M \to Q$ to a set Q is called uniform if for all
$t \in M$, $a \in X$, $A, B \subseteq X$ the following two conditions are satisfied:

i) The value of $\varphi(\partial_{A \cup B}(t))$ is uniquely determined by $\varphi(\partial_A(t))$ and $\varphi(\partial_B(t))$.

ii) The value of $\varphi(\partial_{D(a)}(ta))$ is uniquely determined by $\varphi(\partial_{D(a)}(t))$ and the
letter a.

Our definition of a uniform mapping differs slightly by the one given in
[CM87] since we additionally demand ii). Note also that a uniform mapping
$\varphi : M \to Q$ gives, in general, no natural M-automaton structure to Q; the
reason is that uniformity is in general not enough to calculate the value of $\varphi(ta)$
from $\varphi(t)$ and a if $t \neq \partial_{D(a)}(t)$. However, the point is that a uniform mapping
$\varphi : M \to Q$ induces a natural structure of an asynchronous cellular automaton
on the cartesian product $Q^X = \prod_{a \in X} Q_a$ with $Q_a = Q$ for all $a \in X$. This is
expressed in the next proposition.

Proposition 2.4.4 Let $\varphi : M \to Q$ be a uniform mapping to a set Q and $L \subseteq Q$
be a subset, $M = M(X, D)$. Then there is an asynchronous cellular automaton
with state set Q^X which accepts the trace language $\varphi^{-1}(L)$. In particular, if Q is

finite then $\varphi^{-1}(L)$ is recognizable and accepted by a finite asynchronous cellular automaton, and hence by a finite asynchronous automaton.

Proof: We construct an asynchronous cellular automaton $U = (Q^X, \delta, q_0, F)$ which accepts $\varphi^{-1}(L)$ as follows. The domain of the partially defined transition mapping $\delta : Q^X \times M \to Q^X$ is the set $Z \times M$ where $Z = \{(\varphi(\partial_x(t)))_{x \in X} \mid t \in M\}$. For $(\varphi(\partial_x(t)))_{x \in X}$ and $a \in X$ we set $\delta((\varphi(\partial_x(t)))_{x \in X}, a) = (\varphi(\partial_x(ta)))_{x \in X}$. Since $\partial_x(ta) = \partial_x(t)$ for all $x \neq a$, we see that the action $a \in X$ changes the a-component of $(\varphi(\partial_x(t)))_{x \in X}$ only. Furthermore, the value of $\varphi(\partial_a(ta))$ is uniquely determined by $\varphi(\partial_{D(a)}(t))$ and a by condition ii) of a uniform mapping and $\varphi(\partial_{D(a)}(t))$ is known from $(\varphi(\partial_b(t)))_{b \in D(a)}$ by condition i). In particular $\delta : Z \times M \to Z$ is well-defined and it satisfies the requirement for the partially defined transition mapping of asynchronous cellular automata. Finally we put $q_0 = (\varphi(1), \ldots, \varphi(1))$ and $F = \{(\varphi(\partial_x(t)))_{x \in X} \mid \varphi(t) \in L\}$. To see that the asynchronous cellular automaton U accepts $\varphi^{-1}(L)$ it is enough to notice that the mapping $\varphi : M \to Q$ factorizes through the surjective mapping $\psi : M \to Z$, $\psi(t) = (\varphi(\partial_x(t)))_{x \in X}$. Indeed, condition i) asserts that $\xi : Z \to Q \; \xi(\psi(t)) = \varphi(t)$ is well-defined. Together with Proposition 2.4.3 we obtain the result. $\square$

By Proposition 2.4.4 the strategy to prove Zielonka's Theorem 2.4.2 will be to show that every homomorphism $h : M \to N$ to a (finite) monoid N factorizes through a uniform mapping $\varphi : M \to Q$ to a (finite) set Q.

In the following we write $\partial_{A,B}(t)$ as a short hand for $\partial_A(\partial_B(t))$.

The second approximation of a trace has been introduced in [CM87], here we give a slightly different definition which is notational simpler but gives at the beginning less information about a trace. Let $t \in M$ be a trace then the second approximation $\Delta_2(t)$ is defined as a directed graph with vertex set $X \times X$ and edge set $\{((a, c), (b, d)) \mid \partial_{a,c}(t) \text{ is a prefix of } \partial_{b,d}(t)\}$. Note that in our definition it may happen the graph $\Delta_2(t)$ does not tell us whether $\partial_{a,b}(t) = 1$. However, later we will use a labelled version of $\Delta_2(t)$ and then a vertex $(a, b) \in X \times X$ will have label zero if and only if $\partial_{a,b}(t) = 1$, therefore this information will be retained later.

Remark 2.4.5 i): Let $A, B, C, D \subseteq X$ be non-empty subsets. Then $\partial_{A,C}(t)$ is a prefix of $\partial_{B,D}(t)$ if and only if for all vertices $(a, c) \in A \times C$ of $\Delta_2(t)$ there exists a directed edge to some $(b, d) \in B \times D$. Thus the graph $\Delta_2(t)$ contains the information about the prefix relation between $\partial_{A,C}(t), \partial_{B,D}(t)$ if $A, B, C, D \neq \emptyset$. ii): In a forthcomming paper of Cori, Métivier and Zielonka (personal communication) it is shown that one can avoid the calculation of the second approximation, thereby simplifying the proof. However, the concept of the second approximation seems to be of independent interest in trace theory. Therefore we think that the "classical" proof is still instructive. $\square$

Lemma 2.4.6 *Let $t \in M$ be a trace and $x \in X$ be a letter. Then the second approximation $\Delta_2(tx)$ is computable from $\Delta_2(t)$ and the letter x.*

Proof:

Let $a, b \in X$ and $D(x) = \{y \in X \mid (x, y) \in D\}$. Then we have the following relations:

$$\begin{aligned}
\partial_{a,b}(tx) &= \partial_{a,b}(t) \text{ if } b \neq x, \\
\partial_{a,b}(tx) &= \partial_{a,D(x)}(t) \text{ if } a \neq b = x, \\
\partial_{a,b}(tx) &= \partial_{D(x)}(t)x \text{ if } a = b = x.
\end{aligned}$$

Observe that $\partial_{D(x)}(t)x$ is no prefix of t and that a prefix of t is a prefix of $\partial_{D(x)}(t)x$ if and only if it is a prefix of $\partial_{D(x)}(t)$. Further, $\partial_{D(x)}(t)x$ is a prefix of $\partial_{D(y)}(t)y$ if and only if $x = y$. Hence, we may compute the edge set of $\Delta_2(tx)$ from the relations above using Remark 2.4.5 i). □

Lemma 2.4.7 *Let $t \in M$ be a trace and $A, B \subseteq X$ be subsets such that $t = \partial_A(t) \cup \partial_B(t)$. Write $t_1 = \partial_A(t)$, $t_2 = \partial_B(t)$, $r = t_1 \cap t_2$ with $t_1, t_2, r \in M$, and let $C = \{c \in X \mid \partial_C(t_1) = \partial_C(t_2)\}$. Then we have $r = \partial_C(t) = \partial_C(t_1) = \partial_C(t_2)$. In particular, we have $\max(r) \subseteq C$.*

Proof: Of course, $\partial_C(t) \subseteq r$. For the other direction let $d \in r$ be a fixed point with label $d \in X$. We have $\partial_d(t) = \partial_d(t_1)$ or $\partial_d(t) = \partial_d(t_2)$ since $t = t_1 \cup t_2$. Say $\partial_d(t) = \partial_d(t_1)$ and assume $\partial_d(t) \neq \partial_d(t_2)$. Then $d \notin B$ and there is a path $d < c \leq b$ in t_2 with $d \neq c$, $(d, c) \in D$ and $b \in B$. It follows $c \in r$ and d is not maximal in r. Hence, if $c \in \max(r)$ then $\partial_c(t) = \partial_c(t_1) = \partial_c(t_2)$. Since $r \subseteq \partial_{\max(r)}(t)$ we have $r \subseteq \partial_C(t)$. □

In order to compute the set C from the lemma above we need a labelling of the second approximation. The labelling is based on a function

$$\nu : X \times \mathcal{P}(X) \times M \to \{0, \ldots, \#X\}$$

which is inductively defined as follows:

If $t = 1$ then we put $\nu(a, A, t) = 0$. If we have $\partial_{a,A}(t) \neq t$ then we put $\nu(a, A, t) = \nu(a, a, \partial_{a,A}(t))$. If we have $\partial_{a,A}(t) = t \neq 1$ then we have $a \in A$ and we put $\nu(a, A, t) = \min\{m \geq 0 \mid m \notin \{\nu(a, B, t) \mid a \notin B\}\}$.

It is trivial to see that ν is well-defined as a function to $\mathbb{N}$. The statement $\nu(a, A, t) \leq \#X$ follows from iii) of the lemma below which lists some properties of the function ν which will be used.

Lemma 2.4.8 *We have*

i) $\nu(a, A, t) = \nu(a, a, \partial_{a,A}(t))$ *for all* (a, A, t),

ii) $\nu(a, \emptyset, t) = 0,$

iii) if $A \neq \emptyset$ then $\nu(a, A, t) = \nu(a, b, t)$ for some $b \in A,$

iv) $\nu(a, A, t) = 0$ if and only if $\partial_{a,A}(t) = 1,$

v) if $\partial_{a,A}(t) = \partial_{b,B}(t)$ then $\nu(a, A, t) = \nu(b, B, t).$

Proof: i) and ii) are trivial from the definition. Assertion iii) follows by i) since $\partial_{a,A}(t) = \partial_{a,b}(t)$ for some $b \in A$ if $A \neq \emptyset$. Assertion iv) follows by induction using i), ii) and the fact $a \notin \emptyset$. To see v) observe that $1 \neq \partial_{a,A}(t) = \partial_{b,B}(t)$ implies $a = b$, then use i) and iv). $\square$

The function $\nu : X \times \mathcal{P}(X) \times M \to \{0, \ldots, \#X\}$ induces a function $\nu_t : \Delta_2(t) \to \{0, \ldots, \#X\}$ by $\nu_t(a, b) = \nu(a, b, t)$. We may view ν_t as a labelling function and in the sequel it is this labelled version of the second approximation of a trace which will be used. Using the labelling we may test whether $\partial_{a,b}(t) = 1$ since assertion iv) of the lemma above says $\{(a, b) \in X \times X \mid \partial_{a,b}(t) = 1\} = \{(a, b) \in X \times X \mid \nu_t(a, b) = 0\}$. Note also that assertion v) of the lemma above expresses that the function ν_t is constant on strongly connected components of $\Delta_2(t)$. Our aim is to show that associating with a trace $t \in M$ the pair $(\Delta_2(t), \nu_t)$ yields a uniform mapping. A step in this direction is the following lemma.

Lemma 2.4.9 *Let $t \in M$ be a trace and $x \in X$ be a letter such that $t = \partial_{D(x)}(t)$. Then $(\Delta_2(tx), \nu_{tx})$ is computable from $(\Delta_2(t), \nu_t)$ and $x \in X$.*

Proof: In Lemma 2.4.6 we already have shown how to compute $\Delta_2(tx)$ without the assumption $t = \partial_{D(x)}(t)$. This assumption is used for the computation of ν_{tx}. Let $(a, b) \in X \times X$. Then $\nu_{tx}(a, b) = \nu(a, b, tx) = \nu(a, a, \partial_{a,b}(tx))$. Now, if $b \neq x$ then $\partial_{a,b}(tx) = \partial_{a,b}(t)$. Hence $\nu_{tx}(a, b) = \nu_t(a, b)$ for $b \neq x$. If $a \neq b = x$ then $\nu(a, a, \partial_{a,b}(tx)) = \nu(a, a, \partial_{a,D(x)}(t)) = \nu(a, a, \partial_a(t)) = \nu_t(a, a)$. Here, the use of $\partial_{D(x)}(t) = t$ could have been avoided, but this is not possible in last case where $a = b = x$. Then we have $\nu_{tx}(a, b) = \nu_{tx}(x, x) = \nu(x, x, tx) = \min\{m \geq 0 \mid m \notin \nu(x, B, t), x \notin B\}$ since our assumption implies $\partial_x(tx) = tx$ and $\nu(x, B, tx) = \nu(x, B, t)$ for $x \notin B$. But the values of $\nu(x, B, t)$ for $x \notin B \neq \emptyset$ can be read from the function $\nu_t : \Delta_2(t) \to \{0, \ldots, \#X\}$ since for $B \neq \emptyset$ we find some $b \in B$ such that $\partial_{x,B}(t) = \partial_{x,b}(t)$. $\square$

Proposition 2.4.10 *Let $t \in M$ be a trace and $A, B \subseteq X$ be subsets such that $t = \partial_A(t) \cup \partial_B(t)$. Write $t_1 = \partial_A(t)$, $t_2 = \partial_B(t)$, $r = t_1 \cap t_2$, $t_1 = ru$, $t_2 = rv$ with $t_1, t_2, r, u, v \in M$, $(u, v) \in I$. Let $\varphi(t_i) = (\Delta_2(t_i), \nu_{t_i})$ for $i = 1, 2$. Then the following values are computable from $\varphi(t_1), \varphi(t_2)$:*

i) the set $C = \{c \in X \mid \partial_c(t_1) = \partial_c(t_2)\}$, *i.e., the maximal subset* $C \subseteq X$ *such that* $r = \partial_C(t)$,

ii) for $F \subseteq X$ the set $\mathrm{alph}(\partial_F(v))$,

iii) the second approximation $\Delta_2(t)$,

iv) the function $\nu_t : \Delta_2(t) \to \{0, \ldots, \#X\}$

Proof: i) We show $C = \{c \in X \mid \nu(c, c, t_1) = \nu(c, c, t_2)\}$. Clearly, if $c \in C$, i.e., $\partial_c(t_1) = \partial_c(t_2)$, then

$$\nu(c, c, t_1) = \nu(c, c, \partial_c(t_1)) = \nu(c, c, \partial_c(t_2)) = \nu(c, c, t_2).$$

Now, let $\partial_d(t_1) \neq \partial_d(t_2)$. Without restriction we may assume that $\partial_d(t_1)$ is a prefix of $\partial_d(t_2)$. In particular we have $\partial_d(t_1) \subseteq r$, and, since $\partial_d(t_1) \neq \partial_d(t_2)$, we have $d \notin C$. By Lemma 2.4.7 there must be some $c \in C$ such that $\partial_d(t_1)$ is a prefix of $\partial_c(t_1) = \partial_c(t_2)$ and $\partial_c(t_2)$ is a prefix of $\partial_d(t_2)$. Let $t_2' = \partial_d(t_2)$ then we have $\partial_d(t_1) = \partial_{d,c}(t_2')$ and $\nu(d, d, t_2) = \nu(d, d, t_2') = \min\{m \geq 0 \mid m \notin \{\nu(d, B, t_2') \mid d \notin B\}\}$. Since $d \neq c$ we have $\nu(d, d, t_1) = \nu(d, d, \partial_d(t_1)) = \nu(d, d, \partial_{d,c}(t_2')) = \nu(d, c, t_2') \neq \nu(d, d, t_2') = \nu(d, d, t_2)$.

To prove ii), iii), iv) let us call for a moment $(e, f) \in X \times X$ a u-point if $\partial_{e,C}(t_1)$ is a prefix of $\partial_{e,f}(t_1)$ but $\partial_{e,C}(t_1) \neq \partial_{e,f}(t_1)$, a v-point if $\partial_{e,C}(t_2)$ is a prefix of $\partial_{e,f}(t_2)$ but $\partial_{e,C}(t_1) \neq \partial_{e,f}(t_1)$, and an r-point otherwise. Note that there is a mutual exclusion between these three cases. If C is empty then (e, f) is a u-point if $\nu(e, f, t_2) \neq 0$, (e, f) is a v-point if $\nu(e, f, t_2) \neq 0$ and an r-point in the remaining case $\nu(e, f, t_1) = \nu(e, f, t_2) = 0$. If the set C is non-empty then we may use Remark 2.4.5 to compute from $\Delta_2(t_1)$ whether (e, f) is a u-point and from $\Delta_2(t_2)$ whether (e, f) is a v-point. Assertions ii), iii) and iv) are now easily derived:
ii) For $F \subseteq X$ we have

$$\mathrm{alph}(\partial_F(v)) = \{e \in X \mid (e, f) \text{ is a } v\text{-point for some } f \in F\},$$

iii) To compute $\Delta_2(t)$ let $(a, c) \in X \times X$ be an α-point and $(b, c) \in X \times X$ a β-point, $\alpha, \beta \in \{r, u, v\}$. If $\{\alpha, \beta\} \subseteq \{r, u\}$ then $\partial_{a,c}(t)$ is a prefix of $\partial_{b,c}(t)$ if and only if $\partial_{a,c}(t_1)$ is a prefix of $\partial_{b,d}(t_1)$. Analogously, we may use $\Delta_2(t_2)$ for $\{\alpha, \beta\} \subseteq \{r, v\}$. In the remaining case $\{\alpha, \beta\} = \{u, v\}$ there is no prefix relation between $\partial_{a,c}(t)$ and $\partial_{b,d}(t)$.
iv) If $(e, f) \in X \times X$ is an r-point then $\nu(e, f, t) = \nu(e, f, t_1) = \nu(e, f, t_2)$. If $(e, f) \in X \times X$ is a u-point (a v-point respectively) then $\nu(e, f, t) = \nu(e, f, t_1)$ $(= \nu(e, f, t_2)$ respectively). $\square$

Corollary 2.4.11 *Let Q be the finite set $Q = \{(\Delta_2(t), \nu_t) \mid t \in M\}$. Then the mapping $\varphi : M \to Q$, $\varphi(t) = (\Delta_2(t), \nu_t)$ is uniform.*

Proof: This follows from i), ii), iv) of the proposition above together with Lemma 2.4.9. $\square$

To complete the proof of Theorem 2.4.2 we use the Zielonka's ∇-function. Let $E \subseteq X$ be a subset and $\overline{E} = X \setminus E$. For a trace $t \in M$ the trace $\nabla_E(t)$ is defined by the identity $t = \partial_{\overline{E}}(t)\nabla_E(t)$. Thus, $\nabla_E(t)$ is the maximal suffix s of t such that $\mathrm{alph}(s) \subseteq E$.

Lemma 2.4.12 *Let $t \in M$ be a trace, $x \in X$ be a letter and $I(x) = \{y \in X \mid (x,y) \in I\}$. Then it holds $\nabla_E(tx) = \nabla_E(t)x$ if $x \in E$ and $\nabla_E(tx) = \nabla_{E \cap I(x)}(t)$ otherwise.*

Proof: Let $\overline{E} = X \setminus E$. If $x \in E$ then we have $\partial_{\overline{E}}(tx) = \partial_{\overline{E}}(t)$. If $x \notin E$ then we have $\partial_{\overline{E}}(tx) = \partial_{\overline{E} \cup D(x)}(t)x$. $\square$

Lemma 2.4.13 *Let $t \in M$ be a trace and $A, B \subseteq X$ be subsets such that $t_1 = \partial_A(t)$, $t_2 = \partial_B(t)$, $r = t_1 \cap t_2$, $r = \partial_C(t)$, $t_2 = rv$ with $t_1, t_2, r, v \in M$, $C \subseteq X$. Let $E \subseteq X$ be a subset and $F = X \setminus E$ then we have $\nabla_E(t) = \nabla_G(t_1)\nabla_H(t_2)$ for $G = \{e \in E \mid \{e\} \times \mathrm{alph}(\partial_F(v)) \subseteq I\}$ and $H = E \setminus C$.*

Proof: Clearly, $G \cup H \subseteq E$. It is easy to see that $\nabla_H(t_2) = \nabla_E(v)$. To see that $\nabla_G(t_1)\nabla_H(t_2) = \nabla_G(t_1)\nabla_E(v)$ is a suffix of t observe that $\nabla_G(t_1)\partial_F(v) = \partial_F(v)\nabla_G(t_1)$ and $\partial_F(v)\nabla_E(v) = v$. Now, the equality $\nabla_E(t) = \nabla_G(t_1)\nabla_H(t_2)$ follows since every point of $\nabla_E(t)$ which belongs to t_1 must be independent of every point in $\partial_F(v)$. $\square$

Proposition 2.4.14 *Let $h : M \to N$ be a homomorphism of a trace monoid M to a (finite) monoid N and let Q be the (finite) set*

$$Q = \{(\Delta_2(t), \nu_t, (h\nabla_E(t))_{E \subseteq X}) \mid t \in M\}$$

Then the mapping $\varphi : M \to Q$, $\varphi(t) = (\Delta_2(t), \nu_t, (h\nabla_E(t))_{E \subseteq X})$ is uniform and h factorizes through φ.

Proof: In Corollary 2.4.11 we have shown that the mapping is uniform which associates with a trace its labelled second approximation. Therefore the proof of the proposition above follows directly from Lemma 2.4.12 and Lemma 2.4.13 if we can show that the subsets G and H mentioned in Lemma 2.4.13 are computable from $(\Delta_2(t), \nu_t)$. For H this follows from Proposition 2.4.10 i) and for G from Proposition 2.4.10 ii). $\square$

Proof of Theorem 2.4.2: Let $L \subseteq M$ be a (recognizable) trace language. Let N be the (finite) quotient monoid of M by the syntactic congruence of L. Then we have $L = h^{-1}h(L)$ for the canonical homomorphism $h : M \to N$. Let Q be the (finite) set $Q = \{(\Delta_2(t), \nu_t, (h\nabla_E(t))_{E \subseteq X}) \mid t \in M\}$. We have shown in Proposition 2.4.14 that the mapping $\varphi : M \to Q$, $\varphi(t) = (\Delta_2(t), \nu_t, (h\nabla_E(t))_{E \subseteq X})$ is uniform. Since $\nabla_X(t) = t$ for all $t \in M$ we have that $\varphi(t) = \varphi(t')$ implies $h(t) = h(t')$ for all $t, t' \in M$. Hence, we have $L = \varphi^{-1}\varphi(L)$. The existence of a (finite) asynchronous cellular automaton which accepts L follows from Proposition 2.4.4. From the uniform automaton the construction for the (finite) asynchronous automaton accepting L is given in Proposition 2.4.3. $\square$

3. Petri Nets and Synchronizations

3.1 Local Morphisms of Petri Nets

Since the initiating work of A. Mazurkiewicz, [Maz77], traces have been successfully applied to Petri net theory. Originally the application of trace theory was limited to so-called safe nets. For these nets, traces describe the possible concurrency completely, whereas for more general nets, traces describe only that part of concurrency which is given by the static net topology, and not that which is given by the dynamic behavior. The aim of this chapter is to show that traces are an important tool for the investigation of Petri nets in general.

A *Petri net* is a tuple $N = (P, T, F, B)$ where P is a finite set of places, T is a finite set of transitions, F and B are $P \times T$-matrices over $\mathbb{N}$, the forward and backward incidence matrices. We also view F and B as mappings $F, B : T \to \mathbb{N}^P$. By $\mathbb{N}^P$ we denote as usual the set of mappings from P to the non-negative integers. A *marking* of N is an element of $\mathbb{N}^P$. A transition $a \in T$ is *enabled* under a marking $m \in \mathbb{N}^P$ if $m \geq F(a)$. If $a \in T$ is enabled under $m \in \mathbb{N}^P$ then the follower marking $m' = m - F(a) + B(a)$ may occur. We denote this by $m[a\rangle m'$. For a sequence $w = a_1 a_2 \ldots a_n \in T^*$, we write $m[w\rangle m'$ if $m[a_1\rangle m_1[a_2\rangle m_2 \ldots m_{n-1}[a_n\rangle m'$ for some $m_1, m_2, \ldots, m_{n-1} \in \mathbb{N}^P$. We also write $m[w\rangle$ to denote that $m[w\rangle m'$ holds for some $m' \in \mathbb{N}^P$. If we have $m[w\rangle m'$ then we call m' a *follower marking* of m. Note that according to this definition every marking is a follower marking of itself. We say that two transitions $a, b \in T$ are *concurrently enabled* under $m \in \mathbb{N}^P$ if $m \geq F(a) + F(b)$.

If a Petri net is viewed as a graph then $F(p, a)$, if it is nonzero, denotes the weight of an arc from p to a, if $F(p, a)$ is zero then there is no such arc, and analogously a nonzero value of $B(p, a)$ is the weight of an arc from a to p. Often, a net is given with an initial marking $m_0 \in \mathbb{N}^P$. Then we shall speak of a *system*. If N is a system then $L(N) = \{w \in T^* \mid m_0[w\rangle\}$ denotes the language of N. A system is called *1-safe* (or simply *safe*) if $m(p) \leq 1$ for all $p \in P$ and all follower markings of m_0.

We do not use capacities. They can be handled by introduction of complementary places.

For concurrent systems a partial order semantics is especially suitable. One way to give such a semantics to a net is to associate a dependence alphabet with it. For 1-safe net systems this was first done in [Maz77] and has been taken up by many people. For arbitrary Petri nets $N = (P, T, F, B)$ we associate a dependence alphabet as follows. We say that transition $a, b \in T$ are dependent if

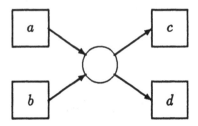

Figure 3.1. Transitions a, b as well as c, d are independent

$a = b$ or a, b are adjacent to some common place p such that $F(p, a) \cdot B(p, b) \neq 0$ or $B(p, a) \cdot F(p, b) \neq 0$. Thus, two different transitions a, b are dependent if and only if at some place of the net there is an arc from a to this place and from this place to b or vice versa. The dependence relation of N is denoted by $D(N)$, its complement $I(N) = T \times T \setminus D(N)$ is called the independence relation of N. Thus in Figure 3.1 we have $(a, c), (a, d), (b, c), (b, d) \in D(N)$ and $(a, b), (c, d) \in I(N)$. With a Petri net N we henceforth associate the dependence alphabet $(T, D(N))$ and the trace monoid $M(N) = M(T, D(N))$.

Our independence relation can also be characterized as follows. Two different transitions $a, b \in T$ are independent if and only if a, b are concurrently enabled under all markings at which some occurrence sequence may begin with ab or ba. In particular, if $w \in T^*$ is an occurrence sequence and $w' \in T^*$ denotes the same trace in $M(N)$ then w' is also enabled. It is therefore sensible to speak of a trace enabled under a marking. Consequently, we associate with a system $N = (P, T, F, B; m_0)$ the trace language $L_\Theta(N) = \{t \in M(N) \mid m_0[t\rangle\}$.

In [Maz77] or [Maz87] a dependence relation for 1-safe nets is defined by the property of being in the neighbourhood of a common place. Thus in Figure 3.1 the transitions a, b, c, d would be pairwise dependent according to Mazurkiewicz' definition. However for 1-safe nets both definitions of dependence are not essentially different. If two transitions a and b are independent for us, but dependent in [Maz87], then in a 1-safe net there can be no occurrence sequence $wabv$ or $wbav$. This results simply from the fact that in a 1-safe net two transitions are never concurrently enabled if they are in the neighbourhood of a common place. Therefore the trace languages associated to a 1-safe system here and in [Maz87] can be identified. Trace languages of 1-safe systems are always recognizable. The next proposition implies, as a special case, that the star of such a language is still recognizable.

Proposition 3.1.1 *Let $N = (P, T, F, B)$ be a Petri net with some initial marking. If $L_\Theta(N)$ is recognizable in $M(N)$ then $L_\Theta(N)^*$ is also a recognizable trace language.*

Proof: Trace languages of a system are always prefix closed. Hence the proof of the propositon follows from Corollary 2.2.7 and Remark 2.2.8 i). $\square$

We now introduce a new type of morphism between nets which we call local morphism. All results in this sections will be based on this kind of morphism.

Definition: *A local morphism* $h : N' \to N$ *from a Petri net* $N' = (P', T', F', B')$ *to a Petri net* $N = (P, T, F, B)$ *is a pair of mappings* $h = (h_P, h_T)$ *such that* $h_P : P' \to P$ *and* $h_T : T' \to T$ *satisfy the following properties:*

i) *for all* $a', b' \in D(N')$, $a' \neq b'$ *it holds* $h_T(a') \neq h_T(b')$,

ii) *for all* $a \in T$ *and for all* $p' \in P'$ *we have*

$$\sum_{a' \in h_T^{-1}(a)} F(p', a') = F(h_P(p'), a)$$

and

$$\sum_{a' \in h_T^{-1}(a)} B(p', a') = B(h_P(p'), a).$$

The first condition says that the images of the preset and the postset are disjoint for all places. The second condition expresses the compatibility of the mappings h_P and h_T with the incidence matrices F and B. In particular, forward arcs are mapped to forward arcs and backward arcs are mapped to backward arcs.

Restricted to the neighbourhood of a place, a local morphism looks like in Figure 3.2. And vice versa: if locally at all places a mapping looks like this then it defines a morphism of nets. (Therefore the name.)

An example of a local morphism is given in Figure 3.3 where the local morphism $h : N' \to N$ is indicated by the labelling.

In order to prove that the composition of two local morphisms is a local morphism again, we state the following lemma. Its proof is trivial from ii) above.

Lemma 3.1.2 *Let* $(h_P, h_T) : N' \to N$ *be a local morphism and* $(a', b') \in D(N')$ *be a pair of dependent transitions then* $h_T(a')$ *and* $h_T(b')$ *are dependent, too.* $\square$

Recall that any mapping $f : A \to B$ of finite sets induces a homomorphism $f^* : \mathbb{N}^B \to \mathbb{N}^A$ defined by $f^*(m) = mf : A \xrightarrow{f} B \xrightarrow{m} \mathbb{N}$. Applied to a local morphism $(h_P, h_T) : N' \to N$ we obtain homomorphisms $h_T^* : \mathbb{N}^T \to \mathbb{N}^{T'}$ and $h_P^* : \mathbb{N}^P \to \mathbb{N}^{P'}$. By linear extension we may view the forward and backward incidence matrices also as homomorphism $F, B : \mathbb{N}^T \to \mathbb{N}^P$. Writing the elements of $\mathbb{N}^T$ and $\mathbb{N}^P$ as formal sums $\sum_{a \in T} n_a a$ and $\sum_{p \in P} n_p p$ respectively where $n_a, n_p \in \mathbb{N}$ we then have

$$F(\sum_{a \in T} n_a a) = \sum_{p \in P} (\sum_{a \in T} n_a F(p, a)) p \text{ and}$$
$$B(\sum_{a \in T} n_a a) = \sum_{p \in P} (\sum_{a \in T} n_a B(p, a)) p.$$

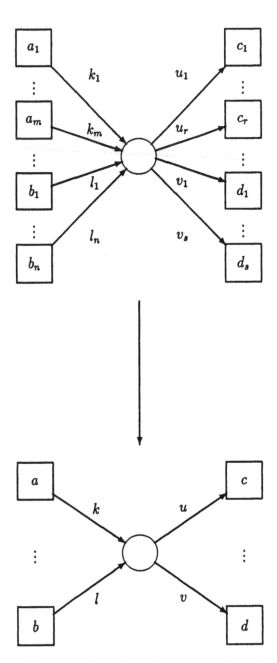

Figure 3.2. A local morphism at a place where the mapping is indicated by the labelling with
$$k = \sum_{i=1}^{m} k_i, \ldots, l = \sum_{i=1}^{n} l_i, u = \sum_{i=1}^{r} u_i, \ldots, v = \sum_{i=1}^{s} v_i.$$

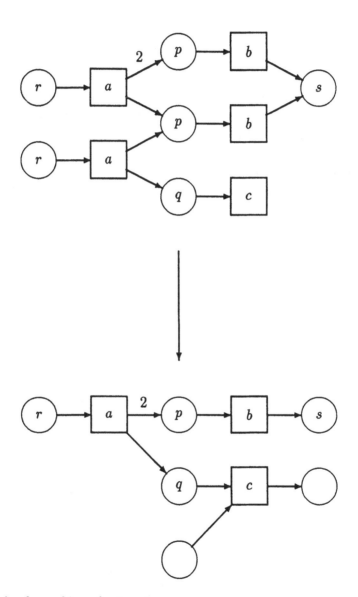

Figure 3.3. A local morphism of nets

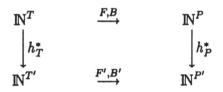

Figure 3.4. Condition ii) of a local morphism

An easy calculation shows that the condition ii) in the definition of a local morphism may be rephrased saying that the diagram of homomorphisms in Figure 3.4 commutes. Together with Lemma 3.1.2 we obtain the following characterization of local morphisms:

Theorem 3.1.3 *Let N' and N be Petri nets and $h_P : P' \to P$, $h_T : T' \to T$ be mappings. Then $(h_P, h_T) : N' \to N$ is a local morphism if and only if the following two conditions are satisfied:*

i') For all $a', b' \in T'$ such that $a' \neq b'$ and $(a', b') \in D(N')$ we have $h_T(a') \neq h_T(b')$ and $(h_T(a'), h_T(b')) \in D(N)$.

ii') The diagram in Figure 3.4 commutes □

Corollary 3.1.4 *Petri nets with local morphisms form a category.* □

From a morphism of nets one would also expect that it relates the behavior of the two respective nets to each other. For local morphisms this works as follows. Condition i') above says that a local morphism $h = (h_P, h_T) : N' \to N$ induces a morphism of undirected graphs between the dependence graphs which underlie the trace monoids $M(N')$ and $M(N)$. Therefore we may define a homomorphism in the opposite direction $h_T^* : M(N) \to M(N')$, which is induced by $h_T^*(a) = \prod_{a' \in h_i^{-1}(a)} a'$ for $a \in T$, see Chap. I, Sect. 1.3. Condition ii') asserts that h_T^* maps the trace language of N to that of N'. This is established in the following two results. From now on we sometimes write h^* instead of h_P^* or h_T^*. It will be clear from the context which mapping is meant.

Corollary 3.1.5 *Let $h : N' \to N$ be a local morphism of Petri nets, $m_1, m_2 \in \mathbb{N}^P$ be markings, and $a \in T$ be a transition of N. Then $m_1[a\rangle m_2$ implies $h^*(m_1)[h^*(a)\rangle\ h^*(m_2)$ where $h^*(a) \in M(N')$ is a step, i.e., $h^*(m_1) \geq F'(h^*(a))$.*

Proof: This is easy by the following sequence of implications:

$$m_1[a\rangle m_2 \Leftrightarrow m_1 \geq F(a) \text{ and } m_2 = m_1 - F(a) + B(a)$$
$$\Rightarrow h^*(m_1) \geq h^*F(a) \text{ and } h^*(m_2) = h^*(m_1) - h^*F(a) + h^*B(a)$$
$$\Leftrightarrow h^*(m_1) \geq F'h^*(a) \text{ and } h^*(m_2) = h^*(m_1) - F'h^*(a) + B'h^*(a)$$
$$\Leftrightarrow h^*(m_1)[h^*(a)\rangle h^*(m_2). \quad \square$$

Corollary 3.1.6 *Let $h : N' \to N$ be a local morphism of Petri nets. Let $m_0 \in$ $\mathbb{N}^P$ be the initial marking of N and $h^*(m_0)$ be the initial marking of N'. Then the homomorphism $h^* : M(N) \to M(N')$ restricts to a mapping*

$$h^* : L_\Theta(N) \to L_\Theta(N') \cap h^*(M(N)).$$

Proof: Iterated application of the corollary above. $\square$

Especially interesting is the case where h^* above is bijective. For this we introduce the following type of local morphisms:

Definition: *A local morphism $(h_P, h_T) : N' \to N$ of Petri nets is called a covering if the mapping h_P is surjective on places and transitions.*

Remark 3.1.7 i) If $h = (h_P, h_T) : N' \to N$ is a covering the h_T induces a morphism on the underlying dependence graphs of $M(N)$ and $M(N')$ which is surjective on vertices and edges.

ii) A local morphism (h_P, h_T) is a covering if h_P is surjective and if every isolated transition is the image of h_T. The surjectivity of h_T follows by property ii) of a local morphism. $\square$

The main theorem of this section is the following:

Theorem 3.1.8 *Let $h = (h_P, h_T) : N' \to N$ be a covering of Petri nets. Let m_0 be the initial marking of N and $h_P^*(m_0)$ be the initial marking of N'. Then the mapping*

$$h_T^* : L_\Theta(N) \to L_\Theta(N') \cap h^*(M(N))$$
$$a_1 \ldots a_n \in T^* \mapsto \left(\prod_{a_1' \in h^{-1}(a_1)} a_1' \right) \ldots \left(\prod_{a_n' \in h^{-1}(a_n)} a_n' \right)$$

is bijective.

Proof: As stated in Remark 3.1.7, h_T viewed as graph a morphism for the respective dependence graphs is surjective on vertices and edges. Thus h^* on $M(N)$ and hence the restriction of h^* to $L_\Theta(N)$ are injective by the general embedding theorem, Corollary 1.4.5.

To see that h^* is surjective let $t \in L_\Theta(N') \cap h^*(\Theta(N))$. We may write t as $h^*(a_1) \ldots h^*(a_m)$ for some $m \geq 0$ and $a_i \in T$ for $1 \leq i \leq m$. Since $h^*(m_0)$

is the initial marking of N' and since $m_1[a\rangle m_2$ implies $h^*(m_1)[h^*(a)\rangle h^*(m_2)$ for all $m_1, m_2 \in \mathbb{N}^P$, $a \in T$, it is enough to show that $h^*(m)[h^*(a)\rangle$ implies $m[a\rangle$ for all $m \in \mathbb{N}^P$, $a \in T$: Now, if we have $h^*(m)[h^*(a)\rangle$ then it holds $h^*(m) \geq \sum_{a' \in h_T^{-1}(a)} F'(a) = F'h^*(a) = h^*F(a)$. Since h_P is surjective on places this implies $m \geq F(a)$ and the result follows. $\square$

The theorem above tells us that if $h : N' \to N$ is a covering then we may split the computation of $L_\Theta(N)$ into two parts: the (local) computation of $L_\Theta(N')$ which might be simpler, and the (global) computation of $h^*(M(N))$ which involves the trace monoid, only. Note that every net has a covering by its atoms and isolated transitions. (An atom is a single place with its adjacent transitions.) Thus, Theorem 3.1.8 has at least one application for every net. But since the computation of the trace language of the covering net becomes so simple in this particular case, difficulties must be hidden somewhere. It is the intersection which is crucial. (This is analogous to the situation of computations with semi-linear sets, see [ES69].) Let us also mention that Theorem 3.1.8 is a formal generalization of the general embedding theorem. This follows, since for every dependence alphabet (X, D) we may construct a net N with $L_\Theta(N) = M(X, D)$ and for every morphism of undirected graphs we obtain a corresponding local morphism of nets.

Another easy application of the theorem above is the case of a local morphism (h_P, h_T) where h_P is surjective and h_T is bijective.

Corollary 3.1.9 *Let $(h_P, h_T) : N' \to N$ be a covering of Petri nets such that h_T is bijective. Let m_0 be the initial marking of N and $h_P^*(m_0)$ be the initial marking of N' and use h_T as an identification of T' and T. Then we have $M(N) = M(N')$ and $L_\Theta(N) = L_\Theta(N')$.*

Proof: Since (h_P, h_T) is covering and h_T is bijective, we may identify the dependence alphabets $(T, D(N))$ and $(T', D(N'))$. The result follows directly from the theorem above. $\square$

Finally note that our results depend substantially on traces: The language of a net may be defined over words, but not over $\mathbb{N}^T$. On the other hand h_T^* has its basic definition over $\mathbb{N}^T$, but h_T^* has no reasonable definition over T^*. It is exactly on the trace monoid where both fits together. Consider also the following commuting diagram:

$$
\begin{array}{ccccccccc}
T^* \supseteq L(N) & \longrightarrow & L_\Theta(N) & \subseteq & M(N) & \longrightarrow & \mathbb{N}^T & \xrightarrow{F,B} & \mathbb{N}^P \\
& & \downarrow h_T^* & & \downarrow h_T^* & & \downarrow h_T^* & & \downarrow h_P^* \\
T^* \supseteq L(N') & \longrightarrow & L_\Theta(N') & \subseteq & M(N') & \longrightarrow & \mathbb{N}^{T'} & \xrightarrow{F',B'} & \mathbb{N}^{P'}
\end{array}
$$

3.2 Synchronization of Petri Nets

A Petri net $N' = (P', T', F', B')$ is called a *subnet* of $N = (P, T, F, B)$ if $S' \subseteq S$, $T' \subseteq T$ and F', B' are the restrictions of F, B. A subnet is called *transition-bounded* if it contains with every place all transitions adjacent to that place. Transition bounded subnets are related to local morphisms as follows:

Lemma 3.2.1 *i) Let* $(h_P, h_T) : N' \to N$ *be a local morphism then* $h_P(P') \subseteq P$ *and* $h_T(T') \subseteq T$ *induce a transition bounded subnet of* N.
ii) Let N' *be any subnet of a net* N *then* N' *is transition bounded if and only if the inclusions* $i_P : P' \hookrightarrow P$, $i_T : T' \hookrightarrow T$ *yield a local morphism of nets* $(i_P, i_T) : N' \to N$.

Proof: i) Let $p \in P$ be a place of N which is adjacent to a transition $a \in T$. If $p = h_P(p')$ for some place $p' \in P'$ then we have $\sum\limits_{a' \in h_T^{-1}(a)} (F(a', p') + B(a', p')) =$
$F(a, p) + B(a, p) > 0$. Hence $h_T^{-1}(a)$ is not empty.
ii) Easy by i). $\square$

Note that if $(i_P, i_T) : N' \to N$ is a local morphism where the mapping i_P is injective then $i_P^* : \mathbb{N}^P \to \mathbb{N}^{P'}$ is just the restriction of a marking of N to a marking of N.

The synchronization of two nets is defined over a common transition bounded subnet. We restrict ourselves to this case in order to have the Theorem 3.2.3 below.

Definition: Let $N_i = (P_i, T_i, F_i, B_i)$, $i = 1, 2$ be Petri nets which contain a common transition bounded subnet $N' = (P', T', F', B')$. Then the synchronization of N_1 and N_2 over N' is defined by $N_1 \|_{N'} N_2 = (P, T, F, B)$ where P and T are the disjoint unions $P = (P_1 \setminus P') \dot\cup (P_2 \setminus P') \dot\cup P'$, $T = (T_1 \setminus T') \dot\cup (T_2 \setminus T') \dot\cup T'$, and F, B are the obvious extensions to P and T. If N_1, N_2 have initial markings m_1, m_2 which coincide on N', then we give $N_1 \|_{N'} N_2$ as initial marking the natural extension of m_1 and m_2 to P.

One should observe that N', N_1 and N_2 may be viewed as transition bounded subnets of the synchronization $N_1 \|_{N'} N_2$. In our category of nets the synchronization is characterized as follows:

Proposition 3.2.2 *The synchronization* $N_1 \|_{N'} N_2$ *of two nets* N_1, N_2 *with a common transition bounded subnet* N' *is a pushout in the category of nets; i.e., for every pair of local morphisms* $h_i : N_i \to N$, $i = 1, 2$ *to a Petri net* N *with* $h_1 i_1 = h_2 i_2 : N' \to N$ *there exists a unique local morphism* $h : N_1 \|_{N'} N_2 \to N$ *such that the diagram in Figure 3.5 commutes:*

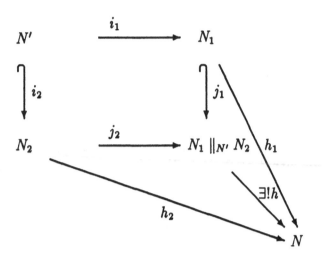

Figure 3.5. The push-out diagram for the synchronization

Proof: Standard □

The synchronization over the empty subnet, which is transition bounded, yields the disjoint union $N_1 \dot\cup N_2$ of nets. This is the direct sum in our category.

If we have two nets N_1, N_2 with a common transition bounded subnet N' then we obtain a natural covering

$$p : N_1 \dot\cup N_2 \rightarrow N_1 \parallel_{N'} N_2.$$

This covering will be used to compute the trace language of a synchronized net. But before we do so, we define the synchronization of traces.

Definition: Let $(X_1, D_1), \ldots, (X_k, D_k)$ be dependence alphabets and let $M_i = M(X_i, D_i)$ be the corresponding free partially commutative monoids, $i = 1, \ldots, k$. Then the synchronization of $M_1, \ldots, M_k$ is defined by

$$M_1 \parallel \ldots \parallel M_2 = M(\bigcup_{i=1}^{k} X_i, \bigcup_{i=1}^{k} D_i).$$

Let $L_i \subseteq M_i$, $i = 1, \ldots, k$. Then the synchronization of $L_1, \ldots, L_k$ is defined by

$$L_1 \parallel \ldots \parallel L_k = \{t \in M_1 \parallel \ldots \parallel M_k \mid p_i(t) \in L_i \text{ for } i = 1, \ldots, k\}$$

where $p_i : M_1 \parallel \ldots \parallel M_k \rightarrow M_i$ denotes the canonical projection, $i = 1, \ldots, k$.

Let us now assume that the net N' is in fact the intersection of nets N_1 and N_2 (we may do this in any case by taking suitable isomorphic copies of N_1 and

N_2), and let us simply write $N_1 \parallel N_2$ in this case. Since N' is transition-bounded, two transitions are dependent in $N_1 \parallel N_2$ if and only if they are dependent in N_1 or in N_2. Thus $M(N_1 \parallel N_2) = M(N_1) \parallel M(N_2)$. Clearly, we can also define the synchronization of several nets $N_1, \ldots, N_k$ with $k \geq 2$. We only have to require that $N_i \cap N_j$ is transition bounded in N_i and N_j for all $i \neq j$. We then set inductively

$$N_1 \parallel \ldots \parallel N_k = (N_1 \parallel \ldots \parallel N_{k-1}) \parallel N_k.$$

The brackets may be omitted since the operation of synchronization is associative. Of course, it is also commutative and $M(N_1 \parallel \ldots \parallel N_k) = M(N_1) \parallel \ldots \parallel M(N_k)$ for the trace monoid of the synchronization of nets $N_1, \ldots N_k$ with $k \geq 2$.

There is a strong compatibility between the synchronization of nets and trace languages. This is stated in the next theorem which turns out to be a special case of Theorem 3.1.8

Theorem 3.2.3 *Let $N_1, \ldots, N_k$ be Petri nets with initial markings, $k \geq 2$, such that for all $1 \leq i, j \leq k$, $i \neq j$ the intersections $P_i \cap P_j$, $T_i \cap T_j$ induce a common transition bounded subnet of N_i and N_j where the initial marking coincides. Then we have*

$$L_\Theta(N_1 \parallel \ldots \parallel N_k) = L_\Theta(N_1) \parallel \ldots \parallel L_\Theta(N_k).$$

Proof: By induction we have to consider the case $k = 2$, only. The natural mapping of the disjoint union to the synchronization, $p : N_1 \dot\cup N_2 \rightarrow N_1 \parallel N_2$, is a covering. Thus, by Theorem 3.1.8, we have $p^*(L_\Theta(N_1 \parallel N_2)) = L_\Theta(N_1 \dot\cup N_2) \cap p^*(M(N_1 \parallel N_2))$. Up to a natural identification we have $M(N_1 \dot\cup N_2) = M(N_1) \times M(N_2)$ and $L_\Theta(N_1 \dot\cup N_2) = L_\Theta(N_1) \times L_\Theta(N_2)$. Since $L_\Theta(N_1) \times L_\Theta(N_2) \cap p^*(M(N_1 \parallel N_2)) = p^*(\{t \in M(N_1 \parallel N_2) \mid p_i(t) \in L_\Theta(N_i) \text{ for } i = 1, 2\}) = p^*(L_\Theta(N_1) \parallel L_\Theta(N_2))$, we obtain $p^*(L_\Theta(N_1 \parallel N_2)) = p^*(L_\Theta(N_1) \parallel L_\Theta(N_2))$. Since p is a covering, the mapping p^* is injective. Hence the result. □

Corollary 3.2.4 *Let $N = (P, T, F, B)$ be a Petri net with initial marking. For $p \in P$ let $\text{Atom}(p)$ the subnet with place p and adjacent transitions. Then we have*

$$L_\Theta(N) = \parallel_{p \in P} L_\Theta(\text{Atom}(p)). \square$$

3.3 Local Checking of Trace Synchronizability

We have seen that the trace language of a synchronized system

$$N_1 \parallel \ldots \parallel N_k$$

can be identified with the set of tuples

$$\{(t_1,\ldots,t_k) \in L_\Theta(N_1) \times \cdots \times L_\Theta(N_k) \mid$$
$$\exists t \in M(N_1 \| \ldots \| N_k) : p_i(t) = t_i \forall i = 1,\ldots,k\}$$

This description of $L_\Theta(N_1 \| \ldots \| N_k)$ contains a global constraint. We have to check the (global) traces $t \in M(N_1 \| \ldots \| N_k)$ whether their projections $p_i(t)$ belong to $L_\Theta(N_i)$ for $i = 1,\ldots,k$. We shall give a graph-theoretical and an algebraic characterization where the synchronization can be described locally. Most of the results are from the revised version of [DV88].

In order to explain what we mean by a local description, we consider the following situation:

Let (X_i, D_i), $i = 1,\ldots,k$ be dependence alphabets with associated monoids $M_i = M(X_i, D_i)$ for $i = 1,\ldots,k$, and the synchronization $M_1 \| \ldots \| M_k = M(\bigcup_{i=1}^{k} X_i, \bigcup_{i=1}^{k} D_i)$. For each $i = 1,\ldots,k$ we denote the canonical projection by $p_i : M_i \| \ldots \| M_k \to M_i$. For $i,j \in \{1,\ldots,k\}$ let $M_{\{i,j\}} = M(X_i \cap X_j, D_i \cap D_j)$ and $p_{ij} : M_j \to M_{\{i,j\}}$ be the canonical mappings. Then we obviously have $p_{ij}p_j = p_{ji}p_i$ for all i,j. Thus, by the general embedding theorem, we obtain a canonical embedding

$$\pi : M_1 \| \ldots \| M_k \to \{(t_1,\ldots,t_k) \in M_1 \times \ldots \times M_k \mid p_{ij}(t_j) = p_{ji}(t_i)$$

for all $i,j\}$

We say that a synchronization $M_1 \| \ldots \| M_k$ has a *local description* if the mapping π above is bijective. The reason is that in this case it can be checked locally on the components of a tuple $(t_1,\ldots,t_k) \in M_1 \times \ldots \times M_k$ whether this tuple represents a trace of $M_1 \| \ldots \| M_k$.

In the sequel we denote the monoid

$$\{(t_1,\ldots,t_k) \in M_1 \times \ldots \times M_k \mid p_{ij}(t_j) = p_{ji}(t_i) \text{ for all } i,j\}$$

by $\bowtie (M_1,\ldots,M_k)$.

Remark 3.3.1 The monoid $\bowtie (M_1,\ldots,M_k)$ can also be defined as a projective limit: For $\emptyset \neq J \subseteq \{1,\ldots,k\}$ let $M_J = M(\bigcap_{j \in J} X_j, \bigcap_{j \in J} D_j)$. We obtain canonical mappings $p_{JK} : M_K \to M_J$ for $\emptyset \neq K \subseteq J \subseteq \{1,\ldots,k\}$ and thereby a projective system $\{(M_K, p_{JK}) \mid \emptyset \neq K \subseteq J \subseteq \{1,\ldots,k\}\}$. It is easy to see that the monoid $\bowtie (M_1,\ldots,M_k)$ is naturally isomorphic to the projective limit of this system. $\square$

For the case $k = 2$ the monoid $\bowtie (M_1, M_2)$ is the usual fibered product and we prefer to denote it by $M_1 \times_{M_{\{1,2\}}} M_2$. But, in general, $\bowtie (M_1, M_2, M_3)$ is not any fibered product of $\bowtie (M_1, M_2)$ and M_3. (In contrast, the synchronization $M_1 \| \ldots \| M_3$ is the synchronization of $(M_1 \| M_2)$ and M_3.)

Here, we are interested in the question in which cases the mapping

$$\pi : M_1 \parallel \ldots \parallel M_k \rightarrow \bowtie (M_1, \ldots, M_k)$$

is an isomorphism.

Similar as in Cori/Métivier [CM85] let us introduce the following notation: If a tuple of traces $(t_1, \ldots, t_k) \in M_1 \times \ldots \times M_k$ belongs to $\pi(M_1 \parallel \ldots \parallel M_k)$ it is called *reconstructible*, if it belongs to $\bowtie (M_1, \ldots, M_k)$ it is called *quasi-reconstructible*. Thus, $\pi : M_1 \parallel \ldots \parallel M_k \rightarrow \bowtie (M_1, \ldots, M_k)$ is an isomorphism if and only if every quasi-reconstructible tuple of traces is reconstructible.

We need a further notion on graphs. Let (V, E) be any (directed) graph. A (directed) cycle in (V, E) is a sequence of different vertices $(x_1, \ldots, x_n)$, $n \geq 3$ ($n \geq 2$) such that there are (directed) edges from x_i to x_{i+1} for all $i \in \mathbf{Z}(\bmod n)$. A cycle is called *chordless* if there are no other edges.

Theorem 3.3.2 *Let as above:*
$M_i = M(X_i, D_i)$ *with dependence graph* G_i, $i = 1, \ldots, k$, $k \geq 2$, $G = \bigcup\limits_{i=1}^{k} G_i$,

$$p_i : M_1 \parallel \ldots \parallel M_k \rightarrow M_i, \quad p_{ji} : M_i \rightarrow M(X_i \cap X_j, D_i \cap D_j)$$

the canonical projections, $1 \leq i, j \leq k$, *and*

$$\bowtie (M_1, \ldots, M_k) = \{(t_1, \ldots, t_k) \in M_1 \times \ldots \times M_k \mid p_{ij}(t_j) = p_{ji}(t_i)$$

for all $i, j\}$.

Then the following assertions are equivalent:

i) *Then canonical embedding*

$$\begin{aligned} \pi : M_1 \parallel \ldots \parallel M_k & \rightarrow & \bowtie (M_1, \ldots, M_k) \\ t & \mapsto & (p_1(t), \ldots, p_k(t)) \end{aligned}$$

   is an isomorphism

ii) *Every chordless cycle in the graph* G *is a cycle in a subgraph* G_i *for some* $i \in \{1, \ldots, k\}$.

Proof: Let $(X, D) = (\bigcup\limits_{i=1}^{k} X_i, \bigcup\limits_{i=1}^{k} D_i)$ and let $\tilde{M}$ be the set of isomorphism classes of finite directed labelled graphs $[V, E, \lambda]$ where V is a finite set of vertices, $E \subseteq V \times V$ is a set of directed edges and $\lambda : V \rightarrow X$ is a labelling function such that the following condition is satisfied for all $x, y \in V$, $x \neq y$:

   It holds $(\lambda(x), \lambda(y)) \in D$ if and only if $(x, y) \in E$ or $(y, x) \in E$.
The set $\tilde{M}$ forms a monoid with the multiplication $[V_1, E_1, \lambda_1][V_2, E_2, \lambda_2]$

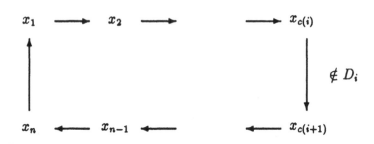

Figure 3.6. A quasi-reconstructible trace which is not reconstructible

$= [V_1 \dot{\cup} V_2, E_1 \dot{\cup} E_2 \dot{\cup} \{(x_1, x_2) \in V_1 \times V_2 \mid (\lambda(x_1), \lambda(x_2)) \in D\}, \lambda_1 \dot{\cup} \lambda_2]$ and the unit element $1 = [\emptyset, \emptyset, \emptyset]$.

Note that $\tilde{M}$ contains $M_1 \parallel \ldots \parallel M_k$ as a submonoid, but it contains also graphs with cycles.

For each $[V, E, \lambda] \in \tilde{M}$ and $i \in \{1, \ldots, k\}$ let $[V, E, \lambda]_i$ be the restriction of the graph to the vertices with labels in X_i and to edges $(x, y) \in E$ with $(\lambda(x), \lambda(y)) \in D_i$. Let $\tilde{\tilde{M}}$ be the submonoid of graphs $[V, E, \lambda]$ such that $[V, E, \lambda]_i$ is acyclic for all $i = 1, \ldots, k$. If $[V, E, \lambda] \in \tilde{\tilde{M}}$ then $[V, E, \lambda]_i$ can be identified with a trace of M_i. Obviously we have $p_{ij}([V, E, \lambda]_j) = p_{ji}([V, E, \lambda]_i)$. Thus, we obtain a homomorphism from $\tilde{\tilde{M}}$ to $\bowtie (M_1, \ldots, M_k)$. One sees easily that this mapping is a bijection. Hence we may identify $\bowtie (M_1, \ldots, M_k)$ with the set of graphs in $\tilde{\tilde{M}}$.

i) $\Rightarrow$ ii): Assume that $(x_1, \ldots, x_n)$ is a chordless cycle in G which is not a cycle in any G_i, $1 \le i \le k$. Then for each $i \in \{1, \ldots, k\}$ there is an index $c(i) \in \mathbf{Z}/(\bmod n)$ such that $(x_{c(i)}, x_{c(i)+1}) \notin D_i$, and this cycle yields a directed labelled graph as in Figure 3.6.

Hence we may identify $(x_1, \ldots, x_n)$ with an element of $\tilde{\tilde{M}} = \bowtie (M_1, \ldots, M_k)$. But this element does not belong $\pi(M_1 \parallel \ldots \parallel M_k)$ since $\pi(M_1 \parallel \ldots \parallel M_k)$ consists of acyclic graphs only. Therefore π is not surjective.

Formally, the directed cycle corresponds to the following element:

$$((x_{c(1)+1} \ldots x_n x_1 \ldots x_{c(1)}), \ldots, (x_{c(n)+1} \ldots x_n x_1 \ldots x_{c(n)}))$$
$$\in \bowtie (M_1, \ldots, M_k) \setminus \pi(M_1 \parallel \ldots \parallel M_k).$$

ii) $\Rightarrow$ i): Assume that $\pi : M_1 \parallel \ldots \parallel M_k \to \bowtie (M_1, \ldots, M_k)$ is not an isomorphism, i.e. π is not surjective. Then there is a graph $[V, E, \lambda]$ in $\tilde{\tilde{M}}$ which contains a directed cycle but the restriction $[V, E, \lambda]_i$ is acyclic for all $i \in \{1, \ldots, k\}$. Let $(x_1, \ldots, x_n)$ be a directed cycle in $[V, E, \lambda]$ of minimal length.

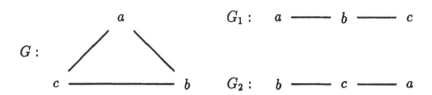

Figure 3.7. G has a chordless cycle which does neither belong to G_1 nor to G_2

We shall show that all labels $\lambda(x_1), \ldots, \lambda(x_n)$ are different. Since for an edge (x, y) of $[V, E, \lambda] \in \tilde{M}$ the pair (y, x) is not an edge, we have $n \geq 3$. Assume that $\lambda(x_i) = \lambda(x_j)$ for different i, j. Since $n \geq 3$ we may assume $i - 1 \not\equiv j \bmod n$. If we had $i + 1 \equiv j \bmod n$ then there are two cases both being impossible: Either $(x_{i-1}, x_j) \in E$ and we find a shorter cycle, or $(x_j, x_{i-1}) \in E$ and we have a triangle with at most two different labels. Thus, all the labels $\lambda(x_1), \ldots, \lambda(x_n)$ are different and $(\lambda(x_1), \ldots, \lambda(x_n))$ is a chordless cycle in the dependence graph G which does not belong to any G_i for $1 \leq i \leq k$. $\square$

Example: Consider a graph as in Figure 3.7. Then G has a chordless cycle, (a, b, c), which neither belongs to G_1 nor to G_2.

Hence $\pi : M_1 \parallel M_2 \to M_1 \times_{M_{\{1,2\}}} M_2$ is not surjective. We have $([abc], [bca]) \in M_1 \times_{M_{\{1,2\}}} M_2$, but $([abc], [bca]) \notin \pi(M_1 \parallel M_2)$. $\square$

On the other hand for the synchronization over a free monoid we have the following positive result:

Corollary 3.3.3 *Let (X_1, D_1), (X_2, D_2) be dependence alphabets and M_1, M_2 be the associated free partially commutative monoids.*

If $M_{\{1,2\}} = M(X_1 \cap X_2, D_1 \cap D_2)$ is free then

$$\pi : M_1 \parallel M_2 \to M_1 \times_{M_{\{1,2\}}} M_2$$

is an isomorphism.

Proof: Let $(x_1, \ldots, x_n)$ be a cycle in G such that $(x_i, x_{i+1}) \notin D_1$ for some $1 \leq i \leq n - 1$ and $(x_n, x_1) \notin D_2$. Then there exists an index $j \in \{1, \ldots, i\}$ and an index $l \in \{i + 1, \ldots, n\}$ such that $x_j, x_l \in X_1 \cap X_2$. If $M_{\{1,2\}}$ is free then we obtain a chord $\{x_j, x_l\}$. $\square$

The corollary above applies to the synchronization of two free monoids. In this particular case it says that the synchronization of two free monoids has always a local description. Such a synchronization corresponds to a covering of the dependence graph by two cliques. For more cliques we may state:

Corollary 3.3.4 *Let G be the dependence graph of a dependence alphabet (X, D). Then there exists a covering by cliques,*

$$(X, D) = (\bigcup_{i=1}^{k} X_i, \bigcup_{i=1}^{k} (X_i \times X_i))$$

such that the resulting embedding

$$\pi : M(X, D) \to\!\!\!\!\bowtie (X_1^*, \ldots, X_n^*)$$

is an isomorphism if and only if every chordless cycle of G is a triangle.

Proof: i) $\Rightarrow$ ii): Every chordless cycle in a complete graph is a triangle.
ii) $\Rightarrow$ i): Every triangle is a clique, and therefore we may choose any covering by cliques such that each triangle lies in at least one of them. $\square$

The corollary above characterizes those trace monoids where we find a covering by cliques such that every quasi-reconstructible tuple of words becomes reconstructible. The next one handles a given covering.

Corollary 3.3.5 *Let $(X, D) = \bigcup_{i=1}^{k} (X_i, X_i \times X_i)$ be a covering of cliques. Then the following assertions are equivalent:*

i) All quasi-reconstructible tuples of $X_1^* \times \ldots \times X_k^*$ are reconstructible.

ii) Let $a_1, \ldots, a_n \in X$, $n \geq 3$. If for all $j \mod n$ there exists some $i \in \{1, \ldots, k\}$ with $a_j, a_{j+1} \in X_i$, then $a_j, a_{j+1}, a_{j+2} \in X_i$ for some $1 \leq j \leq n - 2$ and some $i \in \{1, \ldots, k\}$.

Proof: Let G, G_i be the dependence graphs of (X, D), $(X_i, X_i \times X_i)$, $i = 1, \ldots, k$. Assertion i) means that the canonical embedding π is an isomorphism. Sequences $a_1, \ldots, a_n$ such that for all $j \mod n$ there exists $i \in \{1, \ldots, m\}$ with $a_j, a_{j+1} \in X_i$, as given in ii), are the cycles of G.
i) $\Rightarrow$ ii): By induction on n, the case $n = 3$ being obvious by Theorem 3.3.2. For $n > 3$ the cycle $a_1, \ldots, a_n$ must have a chord $\{a_m, a_l\}$ $m < l$. Applying the induction hypothesis to $a_m, a_{m+1}, \ldots, a_l$ gives the result.
ii) $\Rightarrow$ i): is clear from Theorem 3.3.2. $\square$

Example: Let $M = \{a, b, c, d\}^* / ad = da (\cong \mathbb{N} * \mathbb{N} * \mathbb{N}^2)$ and $X_1 = \{a, b, c\}$, $X_2 = \{b, c, d\}$. Then every quasi-reconstructible pair of words $(w_1, w_2) \in X_1^* \times X_2^*$ is reconstructible. Using the shuffle-operator $\amalg$ of formal language theory we may describe the set of reconstructible pairs as:

$$\{(w_1, w_2) \in X_1^* \times X_2^* \mid \exists v \in \{b, c\}^* : w_1 \in v \amalg a^*, w_2 \in v \amalg d^*\}.$$

Besides the graph theoretical characterization of Theorem 3.3.2 we have the following purely algebraic result.

Theorem 3.3.6 *Use the same notation as in Theorem 3.3.2, but assume furthermore that all alphabets X_i are finite for $i = 1, \ldots, k$. Then the following assertions are equivalent:*

i) The canonical embedding

$$\pi : M_1 \parallel \ldots \parallel M_k \to \bowtie (M_1, \ldots, M_k)$$

is an isomorphism

ii) The monoid $\bowtie (M_1, \ldots, M_k)$ is finitely generated.

Proof: i) $\Rightarrow$ ii): trivial

ii) $\Rightarrow$ i): Assume that $\pi : M_1 \parallel \ldots \parallel M_k \to \bowtie (M_1, \ldots, M_k)$ is not surjective. Then, in the notation of the proof to Theorem 3.3.2, we find a graph in $\tilde{M}$ as in Figure 3.6 above which is a directed cycle and which represents an element of $\bowtie (M_1, \ldots, M_k)$.

For each $m \geq 1$ we may replace x_1 by a power x_1^m and we obtain other elements of $\bowtie (M_1, \ldots, M_k)$ which are represented by directed cycles. Since directed cycles are strongly connected, no such element is a proper product of non-trivial elements. In particular, each strongly connected graph is a necessary generator of $\bowtie (M_1, \ldots, M_k)$. Since we have found infinitely many of them the result follows. $\square$

The previous result shows that the consideration of $\bowtie (M_1, \ldots, M_k)$ leads in a quite natural way to the theory of infinitely generated monoids. However, we are still inside the theory of free partially commutative monoids as we see next.

Theorem 3.3.7 *Let (X_i, D_i), $i = 1, \ldots, k$, $k \geq 2$, be dependence alphabets, $M_i = M(X_i, D_i)$ the associated free partially commutative monoids and $p_{ij} : M_j \to M(X_i \cap X_j, D_i \cap D_j)$, $1 \leq i, j \leq k$, $i \neq j$ be the canonical projections.*
Then the monoid $\bowtie (M_1, \ldots, M_k) = \{(t_1, \ldots, t_k) \in M_1 \times \ldots \times M_k \mid p_{ij}(t_j) = p_{ji}(t_i) \text{ for all } i, j\}$ is free partially commutative.

Proof: Since $\bowtie (M_1, \ldots, M_k)$ is a submonoid of $M_1 \times \ldots \times M_k$ it is enough to show that the Levi Lemma holds in it, see Corollary 1.3.6. For $w \in M_1 \times \ldots \times M_k$ we denote by w_i the corresponding component in M_i and by w_{ij} the element $p_{ij}(w_i) \in M(X_i \cap X_j, D_i \cap D_j)$, $1 \leq i, j \leq k$.

Let $x, y, z, t \in \bowtie (M_1, \ldots, M_k)$ such that $xy = zt$. Then, by the Levi Lemma for M_i, $i = 1, \ldots, k$, we find $r, u, v, s \in M_1 \times \ldots \times M_k$ such that $x = ru$, $y = vs$, $z = rv$, $t = us$, and $uv = vu$ in the direct product $M_1 \times \ldots \times M_k$. Thus, we have to show only that r, u, v, s belong to the submonoid $\bowtie (M_1, \ldots, M_k)$ of $M_1 \times \ldots \times M_k$. By symmetry and the cancellativity it suffices to prove that $r \in \bowtie (M_1, \ldots, M_k)$, i.e., $r_{ij} = r_{ji}$ for all $i \neq j$.

Since $r_{ij} u_{ij} = x_{ij} = x_{ji} = r_{ji} u_{ji}$ for all $i \neq j$, it is enough to show that $|r_{ij}|_a = |r_{ji}|_a$ for all $a \in X_i \cap X_j$, $i \neq j$. Now, for $a \in X_i \cap X_j$, $i \neq j$ we have $|r_{ij}|_a = |r_i|_a = min(|x_i|_a, |z_i|_a) = min(|x_{ij}|_a, |z_{ij}|_a) = min(|x_{ji}|_a, |z_{ji}|_a) = |r_{ji}|_a$. Hence the result. $\Box$

Theorem 3.3.7 is somewhat surprising since the category of free partially commutative monoids is not closed under fibered products, even if we allow letter-to-letter morphisms only. For example, consider the monoid

$$M = \{(w, m, n) \in \{a, b\}^* \times \mathbb{N} \times \mathbb{N} \mid |w| = m + n\}$$

It is easy to see that M is a fibered product of $\{a, b\}^*$ and $\mathbb{N}^2$ without being free partially commutative. Indeed, it results from the equation $(a, 1, 0)(b, 0, 1) = (a, 0, 1)(b, 1, 0)$ that Levi's Lemma does not hold in M.

Putting Theorem 3.2.3 and Theorem 3.3.2 together we obtain a condition such that the trace language of a synchronized system $N_1 \parallel \ldots \parallel N_k$ can be computed locally as $\{(t_1, \ldots, t_k) \in L_\Theta(N_1) \times \ldots \times L_\Theta(N_k) \mid p_{ij}(t_j) = p_{ji}(t_i)$ for all $i \neq j\}$.

We continue with an example where the condition above is satisfied.

Consider two systems N_1, N_2 with intersection N' and synchronization $N_1 \parallel N_2$ as in Figure 3.8.

We have:

$$(T_1, D_1) \;=\; d \overset{a}{\underset{b}{\diamondsuit}} c$$

$$(T_2, D_2) \;=\; d \text{------} c \text{---} e \quad \diagup b$$

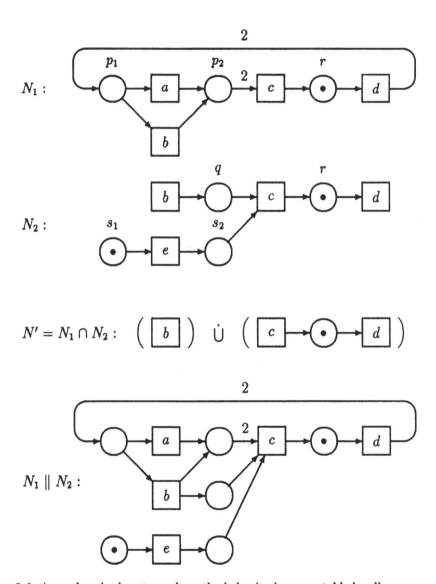

Figure 3.8. A synchronized system where the behavior is computable locally

$$(T_1 \cap T_2, D_1 \cap D_2) \quad = \quad b \;-\!\!-\; c \;-\!\!-\; d$$

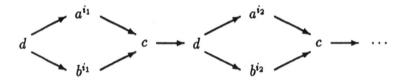

$$(T, D) \quad = \quad$$

Hence $M(N) = M(N_1) \times_{M'} M(N_2)$ where $M' = M(T_1 \cap T_2, D_1 \cap D_2)$ by Theorem 3.3.2. Let us show how to compute $L_\Theta(N) = L_\Theta(N_1 \parallel N_2)$ locally from $L_\Theta(N_1)$ and $L_\Theta(N_2)$. The elements of $L_\Theta(N_1)$ are exactly the prefixes of the following (infinite) traces

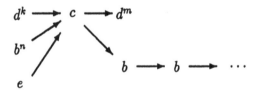

with $2 = (i_1 + j_1) = (i_2 + j_2) = \cdots$
The elements of $L_\Theta(N_2)$ are given by the prefixes of

with $k \leq 1$, $k + m = 2$ and $n \geq 1$.

Let $t_i \in L_\Theta(N_i)$ be traces and $p_i' : M(N_i) \to M' = M(T_1 \cap T_2, D_1 \cap D_2)$ be the canonical projection, $i = 1, 2$. If $t' = p_1'(t_1') = p_2'(t_2')$ then t' is a prefix of

with $n = 1$ or $n = 2$.

This gives us a complete description of $L_\Theta(N)$. We have $t \in L_\Theta(N)$ if and only if t is a prefix of

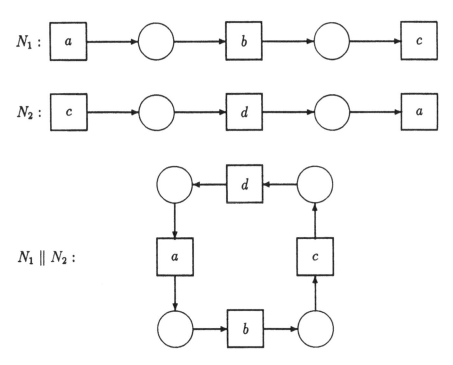

Figure 3.9. A synchronized system where the condition of Theorem 3.3.3 is violated

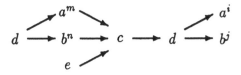

with $m + n = 2$, $n \geq 1$, and $i + j = 2$ or t is a prefix of

$$d \longrightarrow a^2$$
$$e$$

We do not claim that the direct computation of $L_\Theta(N)$ causes any difficulties. The point was that we have determined this behavior purely local from the components. The example was chosen to show how the machinary works only.

If the condition of Theorem 3.3.2 is violated, it may happen that the trace language of a synchronized net system has no local description in the sense above: Consider the situation of systems as in Figure 3.9.

We have

$$[abc] \in L_\Theta(N_1) = \text{Prefix}([(abc)^*]), \text{ and}$$
$$[cda] \in L_\Theta(N_2) = \text{Prefix}([(cda)^*]).$$

Since $p_{21}([abc]) = \{ac, ca\} = p_{12}([cda])$ we have $([abc], [cda]) \in \{(t_1, t_2) \in L_\Theta(N_1) \times L_\Theta(N_2) \mid p_{12}(t_2) = p_{21}(t_1)\}$ but $L_\Theta(N_1 \parallel N_2) = \{1\}$ since there is no token on the net.

This raises the question what can be done if $M(N_1 \parallel \ldots \parallel N_k)$ is not isomorphic to $\bowtie (M_1, \ldots, M_k)$. There is still a local computation if one is willing to renounce some concurrency. Since $L_\Theta(N_1 \parallel \ldots \parallel N_k) = L_\Theta(N_1 \parallel \ldots \parallel N_k) \parallel L_\Theta(N_k)$, it is enough to explain this in the case $k = 2$:

Let (X_1, D_1), (X_2, D_2) be dependence alphabets, $(X', D') = (X_1 \cap X_2, D_1 \cap D_2)$ and $\hat{D}$ be any dependence relation between D' and $X' \times X'$, i.e., $D' \subseteq \hat{D} \subseteq X' \times X'$.

Define: $M_i = M(X_i, D_i)$, $\hat{M}_i = M(X_i, D_i \cup \hat{D})$ for $i = 1, 2$, $M = M(X_1 \cup X_2, D_1 \cup D_2)$, $\hat{M} = M(X_1 \cup X_2, D_1 \cup D_2 \cup \hat{D})$, $M' = M(X', D')$, $\hat{M}' = M(X', \hat{D})$.

We obtain a commutative diagram with canonical mappings:

$$
\begin{array}{ccc}
\hat{M}_1 \parallel \hat{M}_2 & \xrightarrow{\hat{\pi}} & \hat{M}_1 \times_{\hat{M}'} \hat{M}_2 \\
\downarrow{\scriptstyle p} & & \downarrow{\scriptstyle q} \\
M_1 \parallel M_2 & \xrightarrow{\pi} & M_1 \times_{M'} M_2
\end{array}
$$

The canonical projection p is always surjective, but it may happen that q is neither surjective nor injective. Furthermore, it may happen that π is bijective but $\hat{\pi}$ is not, or conversely.

However, we claim that for given $(X_1, D_1), (X_2, D_2)$ there is always at least one choice for $\hat{D} \subseteq X' \times X'$ such that $\hat{\pi} : \hat{M}_1 \parallel \hat{M}_2 \to \hat{M}_1 \times_{\hat{M}'} \hat{M}_2$ becomes an isomorphism. Indeed, in the worst case take $\hat{D} = X' \times X'$ then $\hat{M}'$ is free and we may apply Corollary 3.3.3.

Since for $i = 1, 2$ every trace $t_i \in M_i$ may be viewed as a congruence class of traces in $\hat{M}_i$ it makes sense to write $\hat{t}_i \in t_i$ for $\hat{t}_i \in \hat{M}_i$, $t_i \in M_i$ if the canonical projection maps $\hat{t}_i$ onto t_i. Now, we can use a local description of $\hat{M}_1 \parallel \hat{M}_2$ to compute every synchronization $L_1 \parallel L_2 \subseteq M_1 \parallel M_2$ locally:

Theorem 3.3.8 *With the notations from above assume that $\hat{\pi} : \hat{M}_1 \parallel \hat{M}_2 \to \hat{M}_1 \times_{\hat{M}'} \hat{M}_2$ is an isomorphism. Let $L_1 \subseteq M_1$, $L_2 \subseteq M_2$ be trace languages then π maps $L_1 \parallel L_2$ bijectively onto the following set of pairs:*

$$\{(t_1, t_2) \in L_1 \times L_2 \mid \exists \hat{t}_1 \in t_1, \hat{t}_2 \in t_2 : \hat{p}_{21}(\hat{t}_1) = \hat{p}_{12}(\hat{t}_2) \text{ in } \hat{M}'\}$$

Proof: Since $p : \hat{M}_1 \parallel \hat{M}_2 \rightarrow M_1 \parallel M_2$ is surjective, we obtain from the diagram above $\pi(L_1 \parallel L_2) = q\hat{\pi}(p^{-1}(L_1 \parallel L_2))$. Let $\hat{L}_i \subseteq \hat{M}_i$ be the inverse image of $L_i \subseteq M_i$, $i = 1, 2$. Then one easily verifies $p^{-1}(L_1 \parallel L_2) = \hat{L}_1 \parallel \hat{L}_2$ and hence $\pi(L_1 \parallel L_2) = q\hat{\pi}(\hat{L}_1 \parallel \hat{L}_2)$.

By assumption $\hat{\pi}$ is bijective, hence $\hat{\pi}(\hat{L}_1 \parallel \hat{L}_2) = \{(\hat{t}_1, \hat{t}_2) \in \hat{L}_1 \times \hat{L}_2 \mid \hat{p}_{21}(\hat{t}_1) = \hat{p}_{12}(\hat{t}_2) \text{ in } \hat{M}'\}$; and applying the mapping q to this set yields the result. $\square$

The theorem above yields a local description of $L_1 \parallel L_2$ which is available in all cases. But this description has the disadvantage that one can not work on traces directly; one has to consider all representatives of a trace $t_i \in M_i$ in $\hat{M}_i$ for $i = 1, 2$.

3.4 Algorithms on Three-Colored Graphs

The graph-theoretical characterization for a local description of a synchronization raises the following problem:

$(*)$ Given an undirected graph as a union of subgraphs, $G = \bigcup_{i=1}^{k} G_i$. Is there a chordless cycle which is not a cycle in G_i for some $1 \le i \le k$?

Obviously, this problem is in *NP*. We shall prove that it is *NP*-complete. To see completeness it is enough to consider certain subproblems. We consider only unions of graphs where for each $i \in \{1, \ldots, k\}$ there is exactly one edge which is not an edge of G_i. Denote this edge by e_i, $i = 1, \ldots, k$. Then we look for a chordless cycle containing all edges $e_1, \ldots, e_k$.

Theorem 3.4.1 *The following problem is NP-complete. Given an undirected graph and a subset of edges $\{e_1, \ldots, e_k\}$. Is there a chordless cycle which contains all $e_1, \ldots, e_k$?*

Proof: We reduce the problem to Satisfiability, $(3 - SAT)$, which is well-known to be *NP*-complete, see e.g. [GJ79, p 46]. Let X be a set of variables, $\overline{X}$ be a disjoint copy of X and $V = X \dot{\cup} \overline{X}$. By a clause we mean here an expression of the form $c = v_1 \vee v_2 \vee v_3$ with $v_1, v_2, v_3 \in V$. To each clause c we associate a graph $G(c)$ as in Figure 3.10.

Now, let $e = c_1 \wedge \ldots \wedge c_k$ be an expression where c_i are clauses, $i = 1, \ldots, k$. To the expression e we associate a graph $G(e)$ in the following way:

First we connect each $q_i \in G(c_i)$ with $p_{i+1} \in G(c_{i+1})$ for $i \in \mathbf{Z}(\text{ mod } k)$, and then we draw edges between all $v_i \in G(c_i)$, $v_j \in G(c_j)$ where $1 \le i, j \le k \ i \ne j$ and $\{v_i, v_j\} = \{x, \overline{x}\}$ for some $x \in X$. (For example, the expression

$$(x \vee y \vee z) \wedge (\overline{x} \vee y \vee z) \wedge (x \vee y \vee \overline{z})$$

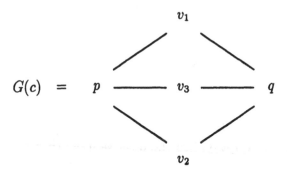

$$G(c) \;=\; p \;—\; v_3 \;—\; q$$

Figure 3.10. The graph associated to a clause

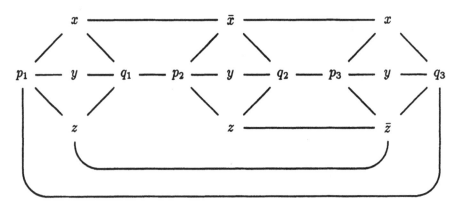

Figure 3.11. The graph of $(x \vee y \vee z) \wedge (\bar{x} \vee y \vee z) \wedge (x \vee y \vee \bar{z})$

is associated with the graph given in Figure 3.11.) It is clear from the construction that the expression $e = c_1 \wedge \ldots \wedge c_k$ has a true assignment if and only if $G(e)$ has a chordless cycle containing all edges

$$\{q_1, p_2\}, \{q_2, p_3\}, \ldots, \{q_{k-1}, p_k\}, \{q_k, p_1\}$$

□

The theorem shows that it might be very difficult to decide whether a synchronization $M_1 \parallel \ldots \parallel M_k$ has a local description. In fact, we have shown:

Corollary 3.4.2 *The following problem is Co-NP-complete:*

Given finite dependence alphabets $(X_1, D_1), \ldots, (X_k, D_k)$, $k \geq 2$. *Let* $M_1, \ldots, M_k$ *be the associated trace monoids. Is the canonical embedding*

$$\pi : M_1 \parallel \ldots \parallel M_k \to\bowtie (M_1, \ldots, M_k)$$

surjective? □

In the rest of this section we restrict ourselves to the case $k = 2$, i.e., $G = G_1 \cup G_2$. In this case it is very suggestive to work with colors. We color an edge of G blue if it belongs to G_1 but not to G_2, yellow if it belongs to G_2 but not to G_1, and green if it belongs to $G_1 \cap G_2$. ('Green' stands for the blending of blue and yellow).

Note that every edge has exactly one color.

In terms of colors, a chordless cycle of $G = G_1 \cup G_2$ which lies neither in G_1 nor in G_2 is the same as a chordless cycle in G with some blue edge and some yellow edge.

In the following we call a graph with a coloring of its edges with blue, yellow and green a *three-colored graph*. We denote the respective edge sets by E_{blue}, E_{yellow}, E_{green} and write edges $\{x, y\}$ as xy henceforth.

The problem $(*)$ restricted to the case $k = 2$ becomes:

$(**)$ Given a three-colored graph. Is there a chordless cycle containing some blue edge and some yellow edge?

We can also formulate the following polynomially equivalent problem.

$(***)$ Given a graph and two edges e_1, e_2. Is there a chordless cycle containing e_1 and e_2?

It has recently been shown by K. Reinhardt that $(***)$ is *NP*-complete. This is a strengthening of Theorem 3.4.1 and solves Problem 261 of [Die90b] which has been an open question from [DV88].

If the number of green edges is bounded logarithmically in the number of vertices we can give a polynomial algorithm that answers $(**)$. First we will deal with the case that there are no green edges at all.

We call a chordless cycle with some blue and some yellow edge a *forbidden cycle*.

Algorithm:
function forbidden_cycle_1 (G: three-colored graph
$\qquad\qquad\qquad$ || $G = (V, E)$ has no green edge):boolean;
define a graph H with vertex set $V_{blue} \dot\cup V_{yellow}$, where $V_i = \{X \mid X$ is a connected component of $(V, E_i)\}$ for $i \in \{$ blue, yellow $\}$,
and edge set $\{XY \mid X \in V_{blue}, Y \in V_{yellow}, X \cap Y \neq \emptyset\}$;
return H contains a cycle **or** there are $X \in V_{blue}, Y \in V_{yellow}$ with $|X \cap Y| \geq 2$
endfunction

Theorem 3.4.3 *Given a three-colored graph G without green edges. Then for-bidden_cycle_1 answers (∗∗) in polynomial time.*

Proof: The time bound is obvious, thus we only have to prove correctness. Assume that there are forbidden cycles in G. Take one cycle in such a way that if we walk around the cycle the number of color changes is as small as possible. If this number is two we have found a blue and a yellow component with two vertices in common. Otherwise, since each blue and each yellow component is passed at most once and at each color change we walk from a blue component into a yellow one via a common vertex (and vice versa), we get a cycle in H.

Now assume that forbidden_cycle_1(G) is true. Then both parts of the hypothesis give us the existence of cycles in G which contain some blue and some yellow edge. A chord of such a cycle gives us two shorter cycles one of which contains some yellow and some blue edge. Thus, a shortest cycle in G which contains some blue and some yellow edge is chordless. □

Now we come to an algorithm for three-colored graphs:

For a graph $G = (V, E)$ and vertices $x_1, \ldots, x_n \in V$ let $G - \{x_1, \ldots, x_n\}$ be defined in the obvious way as the graph obtained from G by deleting $x_1, \ldots, x_n$ and their incident edges. For an edge xy that does not lie on any triangle let G/xy be defined in the obvious way as the graph obtained from G by contracting xy, i.e., by eliminating xy and identifying x and y. Since xy does not lie on a triangle it is obvious how to color the edges of G/xy using the coloring of G.

Algorithm:
function forbidden_cycle (G: three-colored graph ‖ $G = (V, E)$):boolean;
if there is some green edge xy
then
return there is some v with $vx \in E_i$ and $vy \in E_j$, $\{i, j\} = \{$ blue, yellow $\}$
("$\{v, x, y\}$ is a forbidden triangle.")
    or forbidden_cycle ($G - \{x\}$)
("vertex x is not involved in a forbidden cycle.")
    or forbidden_cycle ($G - \{y\}$)
("vertex y is not involved in a forbidden cycle.")
    or forbidden_cycle (($G - \{v \mid (x, v, y)$ is a triangle $\})/xy$)
("There are no forbidden triangles with edge xy.")
else
return forbidden_cycle_1(G)
endif
endfunction

Theorem 3.4.4 *There exists a polynomial p such that algorithm forbidden_cycle solves Problem (**) on input $G = (V, E)$ in time $O(3^{\min(|V|,|E_{green}|)} \cdot p(|V|))$.*

Proof: To prove correctness we only have to consider the then-case, which can easily be checked distinguishing several cases. For the time complexity observe that as long as there are green edges we reduce the problem in polynomial time to three subproblems with fewer vertices and fewer green edges.

Now the result follows with Theorem 3.4.3. $\square$

Corollary 3.4.5 *(i) Problem (**) can be solved in polynomial time if the number of green edges is bounded logarithmically in the number of vertices.*
(ii) For dependence alphabets (X_1, D_1), (X_2, D_2), Condition (ii) of Theorem 3.3.2 can be checked in polynomial time if $|(D_1 \cap D_2) \setminus ((X_1 \cap X_2) \times (X_1 \cap X_2))|$ is bounded logarithmically in the number $|X_1 \cup X_2|$, (in particular, if it is bounded by some constant). $\square$

4. Complete Semi-Thue Systems and Möbius Functions

4.1 Semi-Thue Systems

Most of the material of the first section is standard and can be found e.g. in [Jan88]. It serves as a basic for the following. Some new proofs are given if they seemed to be simpler than the ones found elsewhere. In particular, we show that confluence of noetherian semi-Thue systems can be decided on so-called minimal critical pairs. This is a special case of a result of Winkler-Buchberger [WB83] on term rewriting systems. It is of practical interest since in many cases fewer pairs have to be considered to test confluence. Surprisingly minimal critical pairs appear in a very natural way in computations on Möbius functions. This will be described in Sect. 4.4.

A *semi-Thue system* is a subset $S \subseteq X^* \times X^*$ where X denotes a (finite) alphabet. The elements of S are called *rules*, here often written in the form $l \Rightarrow r$ or $r \Leftarrow l$ if $(l, r) \in S$. We do not use the notation $l \to r$ for rules since simple arrows are used in the dependence graphs of traces, which later could lead to confusion. A semi-Thue system $S \subseteq X^* \times X^*$ defines a relation by $w \underset{S}{\Longrightarrow} w'$ for $w, w' \in X^*$ if $w = ulu'$ and $w' = uru'$ for some $u, u' \in X^*$, $(l, r) \in S$. By $\underset{S}{\overset{*}{\Longrightarrow}}$ ($\underset{S}{\overset{+}{\Longrightarrow}}$ resp.) we denote the reflexive and transitive (transitive resp.) closure of $\underset{S}{\Longrightarrow}$. The reflexive, symmetric and transitive closure of $\underset{S}{\Longrightarrow}$ is denoted by $\underset{S}{\overset{*}{\Longleftrightarrow}}$. It is a congruence and defines a quotient monoid $X^*/\underset{S}{\overset{*}{\Longleftrightarrow}}$ which is also denoted by X^*/S. Two systems $S, T \subseteq X^* \times X^*$ are called *equivalent* if $X^*/S = X^*/T$. The *word problem* of S is to decide whether two given strings $u, v \in X^*$ are congruent modulo S. A classical result of E. Post, [Pos47], states that the word problem of finite semi-Thue systems is recursively unsolvable. A very simple example of a semi-Thue system without a solvable word problem has been constructed by G.S. Ceitin, [Cei58], over the alphabet $X = \{a, b, c, d, e\}$. It has only the following seven rules:

$$ac \Leftrightarrow ca \quad , \quad ad \Leftrightarrow da$$
$$bc \Leftrightarrow cb \quad , \quad bd \Leftrightarrow db$$
$$ce \Leftrightarrow eca \quad , \quad de \Leftrightarrow edb$$
$$c^2 a \Leftrightarrow c^2 ae.$$

A semi-Thue system $S \subseteq X^* \times X^*$ is called *noetherian* if there are no infinite chains $w_0 \underset{S}{\Longrightarrow} w_1 \underset{S}{\Longrightarrow} w_2 \underset{S}{\Longrightarrow} w_3 \underset{S}{\Longrightarrow} \dots .$

It is called *confluent* if for all $u, v, w \in X^*$ with $u \underset{S}{\overset{*}{\Longleftarrow}} v \underset{S}{\overset{*}{\Longrightarrow}} w$ there exists a word $z \in X^*$ with $u \underset{S}{\overset{*}{\Longrightarrow}} z \underset{S}{\overset{*}{\Longleftarrow}} w$. We shall also say that a pair (u, w) is confluent if there exists $z \in X^*$ with $u \underset{S}{\overset{*}{\Longrightarrow}} z \underset{S}{\overset{*}{\Longleftarrow}} w$.

A system $S \subseteq X^* \times X^*$ is called *locally confluent* if for all $u, v, w \in X^*$ with $u \underset{S}{\Longleftarrow} v \underset{S}{\Longrightarrow} w$ the pair (u, w) is confluent. Using noetherian induction one can show that a noetherian system is confluent if and only if it is locally confluent. This fundamental result is due to M. Newman, [New42], see also [Hue80]. A semi-Thue system which is both, noetherian and (locally) confluent, is called *complete*. By $\mathrm{Irr}(S)$ we denote the set of *irreducible* words, i.e., the set of words which are not of the form ulv for any $u, v \in X^*$ and $(l, r) \in S$. If a system $S \subseteq X^* \times X^*$ is finite and complete then we may decide the question whether two words are concurrent modulo S in the following way:

Given two input strings $u, v \in X^*$, we compute their irreducible descendants $u \underset{S}{\overset{*}{\Longrightarrow}} \hat{u} \in \mathrm{Irr}(S)$, $v \underset{S}{\overset{*}{\Longrightarrow}} \hat{v} \in \mathrm{Irr}(S)$. The computation of $\hat{u}$ and $\hat{v}$ is possible since S is finite and noetherian. Since S is confluent and since confluence is equivalent to the so-called Church-Rosser property, [New42], we have $u \overset{*}{\Longleftrightarrow}_s v$ if and only if $\hat{u} = \hat{v}$ as strings over X. Thus, if we can present a monoid by a finite complete semi-Thue system, then the monoid is finitely presentable and it has a decidable word-problem. For quite a long time it has been a challenging open problem whether the converse holds. The answer is negative due to a result of C. Squier, [Squ87].

If we know that a system $S \subseteq X^* \times X^*$ is noetherian then we can test local confluence and hence confluence by inspecting critical pairs. Critical pairs arise from left-hand sides of rules of the system which are overlapping. Formally they are defined as follows:

One says that (u, v) is a *critical pair* of S if u, v have either the form $u = u_1 r_1$, $v = r_2 v_2$ for some $u_1, v_1 \in X^*$, $(l_1, r_1), (l_2, r_2) \in S$, $u_1 l_1 = l_2 v_2$ and $|u_1| < |l_2|$ or $u = r_1$, $v = v_1 r_2 v_2$ for some $v_1, v_2 \in X^*$, $(l_1, r_1), (l_2, r_2) \in S$ and $l_1 = v_1 l_2 v_2$.

It is well-known (and easy to see) that S is locally confluent if and only if all critical pairs are confluent. For noetherian systems not all critical pairs need to be considered. We show this in two steps: The first one reduces to the case that no left-hand side of a rule is a factor of any other left-hand side. Then we introduce so-called minimal critical pairs, and we prove that one can decide confluence by only using them.

The first step is done as follows: let $S \subseteq X^* \times X^*$ be any given noetherian semi-Thue system. Let $L \subseteq X^*$ be the basis of the ideal $X^* \setminus \mathrm{Irr}(S)$, i.e., L is the uniquely determined minimal set such that $X^* \setminus \mathrm{Irr}(S) = X^* L X^*$. Then for each $l' \in L$ we choose exactly one rule $(l', r') \in S$. This is possible since

$L \subseteq \{l \in X^* \mid (l,r) \in S \text{ for some } r \in X^*\}$. Let $S' \subseteq S$ be the subsystem defined by the chosen rules (l',r'). Now for every $(l,r) \in S \setminus S'$ we take the rule $(l',r') \in S'$ such that we have $l = ul'v$ for some $u,v \in X^*$; then we test the confluence of the pair $(ur'v,r)$. If all these pairs are confluent then S' is an equivalent noetherian subsystem of S, and S is confluent if and only if S' is confluent. Thus, it is sufficient to show how we can test confluence of noetherian systems where no left-hand side of a rule is a factor of any other left-hand side. For these systems we define minimal critical pairs.

Definition: *Let* $S \subseteq X^* \times X^*$ *be a noetherian semi-Thue system such that no left-hand side of a rule is a factor of any other left-hand side. Then* $(u,v) \in X^* \times X^*$ *is called a minimal critical pair if for some* $u_1, v_1 \in X^*$, $(l_1,r_1),(l_2,r_2) \in S$, $u_1 l_1 = l_2 v_2$, $|u_1| < |l_2|$ *we have* $u = u_1 r_1$, $v = r_2 v_2$ *and for all* $u',v' \in X^*$, $(l,r) \in S$ *the equation* $u'lv' = u_1 l_1$ *implies* $v' = 1$ *and* $l = l_1$ *or* $l = l_2$ *and* $u' = 1$.

The definition rephrased in different terms says that minimal critical pairs arise from those words which contain left-hand sides exactly twice, one being a prefix the other one a suffix both having a non-trivial overlapping. To give an example consider the system $S = \{(abc,r_1),(bcd,r_2),(cde,r_3)\}$. The critical pair with respect to $r_1 de \underset{S}{\Longleftarrow} abcde \underset{S}{\Longrightarrow} abr_3$ is not minimal in our sense since bcd is a left-hand side. There are two minimal pairs $(r_1 d, ar_2)$ and $(r_2 e, br_3)$.

More generally, assume that for some $a_1, \ldots, a_n \in X$ we have

$$S = \{a_i \ldots a_n a_1 \ldots a_{i-1} \Rightarrow r_i \mid 1 \leq i \leq n\}.$$

Then the system S has $n \cdot (n-1)$ critical pairs, but only n of them are minimal.

The following result is a special case of the Winkler-Buchberger [WB83] criterium for citrical pairs of term rewriting systems.

Theorem 4.1.1 *Let* $S \subseteq X^* \times X^*$ *be a noetherian semi-Thue system such that no left-hand side of a rule is a factor of any other left-hand side. Then* S *is confluent if and only if* S *is confluent on all minimal critical pairs.*

Proof: Let $\preceq$ be the relation on X^* defined by $y \preceq x$ if $x \overset{*}{\underset{S}{\Longrightarrow}} uyv$ for some $u,v \in X^*$. Then the relation $\prec$ $(= (\preceq \text{ and } \neq))$ is easily seen to be a well-founded partial ordering. Consider a critical pair (u,v) which is not minimal in the sense defined above. Then we find words $t,w,y,z \in X^+$, $x \in X^*$ and rules $(l_1,r_1),(l_2,r_2),(l,r) \in S$ such that $l_1 = twx$, $l = wxy$, $l_2 = xyz$ and $u = r_1 yz$, $v = twr_2$. Let $s = twxyz$ then, using noetherian induction, we may assume that

s is confluent on all words $s' \in X^*$ such that $s \succ s'$. In particular, since $s \succ twxy$ and $s \succ wxyz$, the critical pairs (r_1y, tr) and (rz, wr_2) are confluent. Hence

$$u = r_1yz \overset{*}{\underset{S}{\Longrightarrow}} w_1z \overset{*}{\underset{S}{\Longleftarrow}} trz \overset{*}{\underset{S}{\Longrightarrow}} tw_2 \overset{*}{\underset{S}{\Longleftarrow}} twr_2 = v$$

for some $w_1, w_2 \in X^*$. Since also $s \succ trz$ we may conclude that the pair (w_1z, tw_2) is confluent, too. It follows that (u, v) is confluent. $\square$

The idea of the proof above is identical to the one which shows that noetherian systems are confluent if they are locally confluent. The only difference is that we used a finer well-founded ordering.

Theorem 4.1.1 applies, in particular, to normalized systems. A semi-Thue system is called *normalized* if for all rules $(l, r) \in S$ the right-hand side r is irreducible and the left-hand side is irreducible with respect to $S \setminus \{(l, r)\}$. It is well-known that every semi-Thue system $S \subseteq X^* \times X^*$ is equivalent to some normalized complete semi-Thue system $\hat{S} \subseteq X^* \times X^*$. Furthermore, if S is complete and if we demand $\mathrm{Irr}(S) = \mathrm{Irr}(\hat{S})$ then $\hat{S}$ is uniquely determined. We shall prove this simple fact later in a slightly more general set-up.

Unfortunately, there is no way to decide whether a given finite system $S \subseteq X^* \times X^*$ has an equivalent finite complete one, see [O'D83, Thm. 2.3].

It is also well-known that following three problems are recursively unsolvable: Given a finite semi-Thue system $S \subseteq X^* \times X^*$:

i) Is $S \subseteq X^* \times X^*$ noetherian?

ii) Is $S \subseteq X^* \times X^*$ confluent?

iii) Is $S \subseteq X^* \times X^*$ locally confluent?

The undecidability of the first problem follows by simulating a Turing machine, see [HL78] for details. A sketch of proof for the unsolvability of ii), iii) may be found in [BO84, Prop. 1.1]. Let us give a simple proof here (it was already reproduced in [Jan88]): Start with any finite symmetric semi-Thue system $S \subseteq X^* \times X^*$ such that X^*/S has an undecidable word problem. For example, take the seven-rule system of Ceitin mentioned above.

Let $x, y, z \notin X$ be new symbols. For $u, v \in X^*$ define a system:

$$T_{u,v} := S \cup \{x \Rightarrow yuz, x \Rightarrow yvz\}.$$

The undecidability of ii) and iii) is immediate from the following claim.

Claim: Let $u, v \in X^*$. Then the following assertion are equivalent

1): $u \overset{*}{\underset{S}{\Longleftrightarrow}} v$,

2): $T_{u,v}$ is confluent,

3): $T_{u,v}$ is locally confluent.

Proof of Claim: 1) $\Rightarrow$ 2): If we have $u \underset{S}{\overset{*}{\Longleftrightarrow}} v$ then $(X \cup \{x, y, z\})^*/T_{u,v}$ is isomorphic to the free product $(X^*/S) * (\{y, z\}^*)$. The implication follows since S is a symmetric system. 2) $\Rightarrow$ 3): trivial. 3) $\Rightarrow$ 1): Since $T_{u,v}$ is locally confluent, the pair (yuz, yvz) is confluent. But this pair is confluent only if $u \underset{S}{\overset{*}{\Longleftrightarrow}} v$. $\square$

The central algorithm for finite noetherian systems is the computation of irreducible descendants. There is a standard algorithm to compute these descendants, which is described below. It uses string variables s and t which may be implemented by two push-down stores. We call it right_reduce since we work on the input string from right to left.

Algorithm:
(Given: alphabet X and $S \subseteq X^* \times X^*$ finite noetherian)
function right_reduce(s:string):string;
var t:string:=1;
while $s \neq 1$ **do**
        $s := \text{lead}(s)$; $t := \text{last}(s)\ t$;
        if $t \in LX^*$ with $L = \{l \in X^* \mid (l, r) \in S\}$
        then $s := sr$, $t := l^{-1}t$ for any $(l, r) \in S$ such that $t \in lX^*$
        endif
endwhile
return t
endfunction

Proof of Correctness: The algorithm terminates since S is noetherian. Let n be the number of executions of the while-loop and let s_i, t_i, $i = 0, \ldots, n$ be the values of the variables s and t after the i-th execution of the loop. Clearly $s = s_0 t_0 \underset{S}{\overset{*}{\Longrightarrow}} s_1 t_1 \underset{S}{\overset{*}{\Longrightarrow}} \ldots \underset{S}{\overset{*}{\Longrightarrow}} s_n t_n = t_n =$ right_reduce(s). It is therefore enough to show that $t_i \in \text{Irr}(S)$ for all $i = 0, \ldots, n$.

Since S is noetherian, we have $t_0 = 1 \in \text{Irr}(S)$. Let $i \geq 1$, then either $t_i = \text{last}(s_{i-1})t_{i-1} \in \text{Irr}(S)$ or t_i is a suffix of t_{i-1}. In the latter case t_i is irreducible by induction. $\square$

R. Book [Boo82, Thm 4.1] has shown that the algorithm above works in linear time for length-reducing systems. Slightly more generally we will see that it works in linear time for weight-reducing systems. These systems are defined as follows:

A *weight function* on X is a function γ from X to the positive integers. This function defines a homomorphism $\gamma : X^* \to \mathbb{N}$ with $\gamma^{-1}(0) = \{1\}$ by summing up the weights. (The weight function $\gamma(x) = 1$ for all $x \in X$ yields just the length of words.) A semi-Thue system $S \subseteq X^* \times X^*$ is called *weight-reducing* if there is a weight function $\gamma : X^* \to \mathbb{N}$ such that $\gamma(l) > \gamma(r)$ for all $(l, r) \in S$.

The reason that we consider weight-reducing systems instead of length-reducing is that the use of weight-reducing systems leads to a larger class of monoids: Let $S = \{ab \to c^2\}$, then S is confluent and weight-reducing, but the monoid $M = \{a, b\}^*/S$ has no presentation by any finite confluent length-reducing semi-Thue system. In fact, one can show that no confluent length-reducing presentation of M exists over any finite alphabet, see [Die87b, Thm. 2.3].

If S is finite then it is decidable whether S is weight-reducing; and if it is then we may effectively compute a weight function $\gamma : X^* \to \mathbb{N}$ such that $\gamma(l) > \gamma(r)$ for all $(l, r) \in S$. Indeed, let $X = \{a_1, \ldots, a_m\}$, $S = \{(l_1, r_1), \ldots, (l_n, r_n)\}$, $m, n \geq 0$. The system S is weight-reducing if and only if the following system of n linear equations with variables $x_1, \ldots, x_m, z_1, \ldots, z_m$ has a strictly positive rational solution:

$$\left(\sum_{i=1}^{m} (|l|_{a_i} - |r|_{a_i}) x_i = z_j \right)_{j=1,\ldots,n}$$

Proposition 4.1.2 ([Boo82, Thm. 4.1]) *If $S \subseteq X^* \times X^*$ is a finite weight-reducing semi-Thue system then algorithm right_reduce works in linear time.*

Proof: Let $\gamma : X^* \to \mathbb{N}$ be a weight function and $0 < \epsilon < 1$ such that $(1 - \epsilon)\gamma(l) \geq \gamma(r) + \epsilon$ for all $(l, r) \in S$. As above, let s be the input, n be the number of while-loops, and s_i, t_i the values of the variables s, t after the i-th loop, $i = 0, \ldots, n$. For $i = 0, \ldots, n$ define $\delta_i := \gamma(s_i) + (1 - \epsilon)\gamma(t_i)$, then we have $0 \leq \delta_i \leq \delta_{i-1} - \epsilon \leq \delta_0 = \gamma(s)$ for $1 \leq i \leq n$. Hence $n \leq \frac{1}{\epsilon\gamma(s)} \in O(|s|)$. $\quad\square$

4.2 Complete Presentations of Trace Monoids

In the last section we have seen that finite, confluent, weight-reducing semi-Thue systems present monoids with a word problem solvable in linear time. The converse is not true. This can be seen by the consideration of $\mathbb{N}^2$. More generally let (X, D) be a finite dependence alphabet with independence relation $I = X \times X \setminus D$. We know that the word problem for traces is solvable in linear time. However, unless $M(X, D)$ is free the monoid $M(X, D)$ is not presentable by any finite confluent weight-reducing semi-Thue system $T \subseteq Y^* \times Y^*$. Indeed assume the contrary. Then for $u, v \in Y^*$ with $u \overset{*}{\underset{T}{\Longleftrightarrow}} v$ there exists a derivation

$u \underset{T}{\Longleftrightarrow} u_1 \underset{T}{\Longleftrightarrow} \ldots \underset{T}{\Longleftrightarrow} u_n = v$ of length $n \in O(|uv|)$. Now this property does not depend on the presentation. It is therefore also satisfied for the system $S = \{ab \leftrightarrow ba \mid (a,b) \in I\}$. If $M(X,D)$ is not free, we find independent letters $a, b \in X$, $(a,b) \in I$. Since every derivation

$$a^m b^m \underset{S}{\Longleftrightarrow} u_1 \underset{S}{\Longleftrightarrow} \ldots \underset{S}{\Longleftrightarrow} u_n = b^m a^m$$

has length at least $n = m^2$, no such $T \subseteq Y^* \times Y^*$ can exist.

On the other hand it is possible to present a trace monoid by some finite complete semi-Thue system without length-increasing rules. Such a complete presentation exists over an alphabet where the letters are the elementary steps of the trace monoid. Recall that the set of elementary steps has been defined by

$$\mathcal{F} = \{F \subseteq X \mid \emptyset \neq F \text{ is finite and } (a,b) \in I \text{ for all } a, b \in F, a \neq b\}$$

Every elementary step $F \in \mathcal{F}$ is associated with a trace $[F] = \prod_{a \in F} a \in M(X,D)$ and identifying an element $a \in X$ with the singleton set $\{a\} \in \mathcal{F}$ we have $X \subseteq \mathcal{F}$.

Theorem 4.2.1 *Let $S \subseteq \mathcal{F}^* \times \mathcal{F}^*$ be the semi-Thue system which has the following rules:*

$$
\begin{aligned}
FF' &\Rightarrow F \cup F' \text{ for } F \cup F' \in \mathcal{F}, \, F \cap F' = \emptyset \\
FF' &\Rightarrow (F \cup F'')(F' \setminus F'') \\
&\quad \text{for } F, F', F'' \in \mathcal{F}, \, \emptyset \neq F'' = \{a \in F' \setminus F \mid F \cup \{a\} \in \mathcal{F}\} \neq F'
\end{aligned}
$$

Then the inclusion $X \subseteq \mathcal{F}$ induces an isomorphism between $M(X,D)$ and $\mathcal{F}^*/S$. The inverse of this isomorphism is given by the mapping $\psi : \mathcal{F} \to M(X,D)$, $F \mapsto [F]$. The semi-Thue system $S \subseteq \mathcal{F}^* \times \mathcal{F}^*$ is complete; it is finite if and only if the independence relation $I \subseteq X \times X$ is finite.

Proof: It is clear that S is finite if and only if I is finite. The system S is noetherian since no rule is length increasing and for each length preserving rule $FF' \Rightarrow (F \cup F'')(F' \setminus F'')$, the set $F' \setminus F''$ is properly included in F'. The confluence of S follows by inspecting the critical pairs. This is easily done and left to the reader. Hence S is complete. To complete the proof, observe that the mapping $\psi : \mathcal{F} \to M(X,D)$ induces a surjective homomorphism $\overline{\psi} : \mathcal{F}^*/S \to M(X,D)$ such that $\overline{\psi}(x) = x$ for all $x \in X$ and where the inverse mapping is induced by the inclusion $X \subseteq \mathcal{F}$. Hence $\overline{\psi}$ is an isomorphism. $\square$

The classical Foata normal form theorem, see Theorem 1.2.1, turns out to be a corollary:

Corollary 4.2.2 *The set of traces in $M(X, D)$ is in canonical bijection with the following set of elementary step sequences:*

$$\{F_1 \ldots F_r \mid r \geq 0, F_i \in \mathcal{F} \text{ for } i = 1, \ldots, r, \quad \forall i = 1, \ldots, r - 1$$
$$\forall b \in F_{i+1} \exists a \in F_i : (a, b) \in D\}$$

Proof: This set is the set of irreducible words with respect to the complete system $S \subseteq \mathcal{F}^* \times \mathcal{F}^*$ given above. $\square$

In order to obtain a finite complete presentation of a trace monoid $M(X, D)$ we used the larger alphabet of elementary steps $\mathcal{F}$. Thus, a natural question arises whether we can avoid this, i.e., whether there is a finite complete system $S \subseteq X^* \times X^*$ which defines $M(X, D)$. In general, no such system can exist by the following result due to F. Otto:

Proposition 4.2.3 ([Ott87]) *Let (X, D) be a finite dependence alphabet with independence relation $I = X \times X \setminus D$. Then there exists a finite complete semi-Thue system $S \subseteq X^* \times X^*$ such that $X^*/S = M(X, D)$ if and only if there is a linear ordering $\leq$ on X such that the relation $\leq \cap I$ is transitive.*
If $\leq \cap I$ is transitive then such system $S \subseteq X^* \times X^*$ is given by the rules:

$$ba \Rightarrow ab \text{ for } a \leq b \text{ and } (a, b) \in I.$$

Proof: Assume first that $\leq \cap I$ is transitive for some linear ordering $\leq$ on X. Clearly, the system $S = \{ba \Rightarrow ab \mid a \leq b, (a, b) \in I\}$ is noetherian, and we have $X^*/S = M(X, D)$. The system S is confluent since every critical pair arises from a situation $cab \underset{S}{\Longleftarrow} cba \underset{S}{\Longrightarrow} bca$, $(a, b), (b, c) \in (\leq \cap I)$ which has a resolution $cab \underset{S}{\overset{*}{\Longrightarrow}} abc \underset{S}{\overset{*}{\Longleftarrow}} bca$ since $(a, c) \in (\leq \cap I)$.

Now, let $S \subseteq X^* \times X^*$ be any finite complete system such that $X^*/S = M(X, D)$. We first show that if $ca \Rightarrow ac \notin S$ then it holds $c^* a^* \subseteq Irr(S)$. Indeed, let $p : X^* \to M(X, D)$ be the canonical projection. If $(a, c) \in D$ then $p^{-1}p(w) = \{w\}$ for every $w \in \{a, c\}^*$. In particular $c^* a^*$ is an irreducible subset of X^*. If $(a, c) \in I$ then $a \neq c$ and $p^{-1}p(ac) = \{ac, ca\}$. Since $ca \Rightarrow ac \notin S$ we must have $ac \Rightarrow ca \in S$. But then every word $w \in \{a, c\}^*$ has a descendant $w \underset{S}{\overset{*}{\Longrightarrow}} w' \in c^* a^*$. Since p restricted to $c^* a^*$ is injective, we conclude that $c^* a^*$ is irreducible in this case, too.

Note that $(a, c) \in I$ implies either $ac \Rightarrow ca \in S$ or $ca \Rightarrow ac \in S$ for all $a, c \in X$. Assume now we would have $ba \Rightarrow ab \in S$, $cb \Rightarrow bc \in S$, but $ca \Rightarrow ac \notin S$. Therefore, we have $a^* b^*, b^* c^*, c^* a^* \subseteq Irr(S)$. Since S is finite we find some large $n \gg 0$ that $Irr(S) \ni c^n a^n b \underset{S}{\overset{*}{\Longleftarrow}} c^n b a^n \underset{S}{\overset{*}{\Longrightarrow}} bc^n a^n \in Irr(S)$. This is a contradiction to the confluence of S. Hence the relation $a < b$, defined by $ba \Rightarrow ab \in S$, is transitive. Now, let $\leq$ be any linear order on X extending $<$ then it holds $< = \leq \cap I$. $\square$

Remark 4.2.4 An analogous result as above on so-called semi-commutation systems was independently obtained by Y. Métivier and E. Ochmanski, see [MO88].

The proposition of A. V. Anisimov and E. Knuth, see 1.2.2, can be stated in terms of infinite complete semi-Thue systems as follows:

Proposition 4.2.5 Let (X, D) be a dependence alphabet, $I \subseteq X^* \times X^*$ the independence relation extended to words and $\leq$ be a linear ordering of X. Then the system $T = \{bua \Rightarrow aub \mid (a, ub) \in I, a < b\}$ is complete and it presents $M(X, D)$. The set $\mathrm{Irr}(T)$ is the set of lexicographic normal forms. $\square$

Putting the last two results together we obtain:

Corollary 4.2.6 Let (X, D) be a finite dependence alphabet and $\leq$ be a linear ordering of X. Then the following assertions are equivalent:

i) The normalization of the semi-Thue system $T \subseteq X^* \times X^*$ above is finite.

ii) The relation $\leq \cap I$ is transitive. $\square$

4.3 Unambiguous Möbius Functions

We first give an informal description of the contents of this section. Let $M = M(X, D)$ be a trace monoid and $\mathbf{Z}\langle\langle M \rangle\rangle$ be the ring of formal power series over M. One of the fundamental properties of a trace monoid concerns the polynomial $\mu_M = \sum (-1)^{|F|}[F] \in \mathbf{Z}\langle\langle M \rangle\rangle$ where the sum is taken over those subsets $F \subseteq X$ which consist of pairwise independent letters only and where $[F]$ denotes the trace $\prod_{x \in F} x \in M$. This polynomial has a formal inverse in $\mathbf{Z}\langle\langle M \rangle\rangle$ which is the constant function with value one. The polynomial $\mu_M \in \mathbf{Z}\langle\langle M \rangle\rangle$ is therefore the so-called Möbius function of M.

If we represent each trace $[F]$ in μ_M by some word $w_F \in X^*$ we obtain a polynomial $\sum (-1)^{|w_F|} w_F$ in the ring of formal power series $\mathbf{Z}\langle\langle X^* \rangle\rangle$. There are several choices for these polynomials, and each such polynomial will be called a *Möbius function* for M. We shall call a Möbius function *unambiguous* if its formal inverse in $\mathbf{Z}\langle\langle X^* \rangle\rangle$ is the characteristic function of a set of representatives for M.

As reported in [Cho86, Chap II.2] it has been conjectured that for every trace monoid there is at least one unambiguous Möbius function. Further it has been asked how to find the words $w_F \in X^*$ such that $\mu = \sum (-1)^{|w_F|} w_F$ is unambiguous. It turns out that the conjecture is false. But we will see that, in case of existence, there is a very simple way to compute these words w_F.

We prove that the existence of unambiguous Möbius functions is directly related to the existence of transitive orientations of I. We have seen in Proposition 4.2.3 that transitive orientations correspond to finite complete semi-Thue systems $S \subseteq X^* \times X^*$ with $X^*/S = M$. We obtain therefore an intimate connection between unambiguous Möbius functions and finite complete semi-Thue systems. We show that the formal inverse of an unambiguous Möbius function is the characteristic series over the rational set of irreducible words of a finite complete semi-Thue system which defines M. Moreover, starting with a finite complete semi-Thue system $S \subseteq X^* \times X^*$ defining M, we obtain, by the formal inverse of the characteristic function over the irreducible words, an unambiguous Möbius function for M. Altogether we obtain a canonical bijection between the following sets: transitive orientations of I, unambiguous Möbius functions for M, finite normalized complete semi-Thue systems which define M.

The existence of a transitive orientation is of purely graph-theoretical nature and can be decided roughly in time $O(\#X^3)$. Thus, there is a polynomial time algorithm which decides for a finite alphabet X with independence relation $I \subseteq X \times X$ whether there is an unambiguous Möbius function in $\mathbf{Z}\langle\langle X^* \rangle\rangle$, and which computes such a function in case of existence.

We now start with the formal discussion. Let (X, D) be a finite dependence alphabet with independence relation $I = X \times X \setminus D$ and associated trace monoid $M = M(X, D)$. An *orientation* of the independence relation I is a subset $I^+ \subseteq I$ such that we have a disjoint union $I = I^+ \dot\cup \{(b, a) \mid (a, b) \in I^+\}$. An orientation I^+ of I is called *transitive* if $(a, b) \in I^+$, $(b, c) \in I^+$ implies $(a, c) \in I^+$ for $a, b, c \in X$. If $I^+ \subseteq I$ is a transitive orientation then X is a partially ordered set, and we will write $a < b$ instead of $(a, b) \in I^+$ for $a, b \in X$.

A mapping from M to $\mathbf{Z}$ is called a *formal power series*, and the collection of all formal power series is denoted by $\mathbf{Z}\langle\langle M \rangle\rangle$. This set has a ring structure by the usual addition

$$(\varphi + \psi)(w) := \varphi(w) + \psi(w)$$

and the multiplication

$$(\varphi \cdot \psi)(w) := \sum_{w_1 w_2 = w} \varphi(w_1)\psi(w_2)$$

for $\varphi, \psi \in \mathbf{Z}\langle\langle M \rangle\rangle$ and traces $w, w_1, w_2 \in M$. The neutral element for the multiplication is the function $\varphi(1) = 1$ and $\varphi(w) = 0$ for all $w \in M$, $w \neq 1$. (For the general theory of formal power series we refer to [KS86].)

A formal power series $\varphi \in \mathbf{Z}\langle\langle M \rangle\rangle$ has a multiplicative inverse φ^{-1}, called the *formal inverse*, if and only if $\varphi(1)$ is a unit in $\mathbf{Z}$, i.e., $\varphi(1) = 1$ or $\varphi(1) = -1$. The formal inverse φ^{-1} is given inductively by $\varphi^{-1}(1) := \varphi(1)$ and for $w \neq 1$ by the identity:

$$-\varphi(1)\varphi^{-1}(w) = \sum_{\substack{w_1 w_2 = w \\ w_1 \neq 1}} \varphi(w_1)\varphi^{-1}(w_2)$$

Every subset $N \subseteq M$ defines a power series $\zeta_N \in \mathbf{Z}\langle\langle M \rangle\rangle$ by its characteristic function $\zeta_N(w) = 1$ if $w \in N$ and $\zeta_N(w) = 0$ otherwise. In particular, ζ_M is the constant function with value 1. Besides in functional notation we also write power series in the form $\varphi = \sum_{w \in M} \varphi(w)w$. This is justified since $M \to \mathbf{Z}\langle\langle M \rangle\rangle$, $w \mapsto \zeta_{\{w\}}$ is an embedding. Then, reading this embedding as an inclusion, the formula $\varphi = \sum_{w \in M} \varphi(w)w$ becomes an identity of $\mathbf{Z}\langle\langle M \rangle\rangle$. The support of a power series $\varphi \in \mathbf{Z}\langle\langle M \rangle\rangle$ is the set $\text{supp}(\varphi) = \{w \in M \mid \varphi(w) \neq 0\}$. Power series with finite support are called *polynomials*.

Consider the finite set $\mathcal{F}_\emptyset := \{F \subseteq X \mid (a, b) \in I \text{ for all } a, b \in F, a \neq b\}$. This definition implies $\emptyset \in \mathcal{F}_\emptyset$. Earlier we have defined the set of elementary steps $\mathcal{F}$. There it has been more convenient to exclude the empty set. Hence $\mathcal{F} = \mathcal{F}_\emptyset \setminus \{\emptyset\}$. Every element $F \in \mathcal{F}_\emptyset$ defines as usual a unique trace $[F] \in M$ by the product $[F] := \prod_{x \in F} x$. The *Möbius function* $\mu_M \in \mathbf{Z}\langle\langle M \rangle\rangle$ is defined by the following polynomial

$$\mu_M := \sum_{F \in \mathcal{F}_\emptyset} (-1)^{|F|}[F]$$

Since $\emptyset \in \mathcal{F}_\emptyset$ and $[\emptyset] = 1$, we have $\mu_M(1) = 1$ and this function has a formal inverse. The most important property of the Möbius function is the following result of Cartier and Foata:

Proposition 4.3.1 ([CF69, Thm. 2.4]) *The formal inverse of the polynomial* $\mu_M = \sum_{F \in \mathcal{F}_\emptyset} (-1)^{|F|}[F]$ *is the constant function with value one:* $\mu_M^{-1} = \zeta_M = \sum_{w \in M} w$. $\square$

Here, we are concerned with the question when we can lift the Möbius function $\mu_M \in \mathbf{Z}\langle\langle M \rangle\rangle$ to a polynomial $\mu \in \mathbf{Z}\langle\langle X^* \rangle\rangle$ such that the formal inverse of μ in $\mathbf{Z}\langle\langle X^* \rangle\rangle$ is a characteristic function over a set of representatives for M.

The link between the rings $\mathbf{Z}\langle\langle X^* \rangle\rangle$ and $\mathbf{Z}\langle\langle M \rangle\rangle$ is given by the canonical projection $p : X^* \to M$. It induces a surjective ring homomorphism, denoted again by:

$$p : \mathbf{Z}\langle\langle X^* \rangle\rangle \to \mathbf{Z}\langle\langle M \rangle\rangle, \qquad p(\varphi) = \sum_{t \in M} \Big(\sum_{p(w)=t} \varphi(w) \Big)t$$

Definition: i): A *Möbius function for M over X* is a power series $\mu \in \mathbf{Z}\langle\langle X^* \rangle\rangle$ such that $p(\mu) = \mu_M$ is the Möbius function in $\mathbf{Z}\langle\langle M \rangle\rangle$ and such that $p : X^* \to M$ is injective on the support of μ.
ii): A Möbius function for M $\mu \in \mathbf{Z}\langle\langle X^* \rangle\rangle$ is called *unambiguous* if its formal inverse in $\mathbf{Z}\langle\langle X^* \rangle\rangle$ is a characteristic function of a set of representatives of M in X^*.

Remark 4.3.2 A Möbius function $\mu \in \mathbf{Z}\langle\langle X^*\rangle\rangle$ is unambiguous if and only if its formal inverse has non-negative coefficients.

Proof: This follows by Proposition 4.3.1 and the definition of the ring homomorphism $p : \mathbf{Z}\langle\langle X^*\rangle\rangle \to \mathbf{Z}\langle\langle M\rangle\rangle$. $\square$

Our ultimative goal is to prove that the set of unambiguous Möbius functions for M is in canonical one-to-one correspondence with the set of transitive orientations of the independence relation I. Moreover, the formal inverse of an unambiguous Möbius function for M will be shown to be the characteristic function over the set of irreducible words of a finite complete semi-Thue system $S \subseteq X^* \times X^*$ which defines the free partially commutative monoid M. In particular, the formal inverse of an unambiguous Möbius function is the characteristic function of a rational cross-section.

Let us start with a transitive orientation I^+ of the independence relation $I \subseteq X \times X$. We write again $a < b$ instead of $(a, b) \in I^+$. We define a finite semi-Thue system $S \subseteq X^* \times X^*$ by:

$$S = S(I^+) = \{(ab, ba) \mid a, b \in X, \ a < b\}$$

As we have seen in the preceeding section $S \subseteq X^* \times X^*$ is a finite complete system defining M.

Let us define a polynomial $\mu = \mu(I^+) \in \mathbf{Z}\langle\langle X^*\rangle\rangle$ by $\mu(w) = (-1)^n$ if $w = a_1 \ldots a_n$, $n \geq 0$, $a_i \in X$ for $1 \leq i \leq n$, $a_i < a_j$ for $1 \leq i < j \leq n$, and put $\mu(w) = 0$ otherwise. We also write $\mu = \sum_{a_1 < \cdots < a_n} (-1)^n a_1 \ldots a_n$.

Theorem 4.3.3 *The polynomial $\mu = \sum_{a_1 < \cdots < a_n} (-1)^n a_1 \ldots a_n$ is an unambiguous Möbius function. Its formal inverse ζ in $\mathbf{Z}\langle\langle X^*\rangle\rangle$ is the characteristic function of the set of irreducible words $Irr(S)$, i.e., $\zeta = \sum_{w \in Irr(S)} w$.*

Proof: Since I^+ is a transitive orientation, the subsets $F \subseteq X$, where $(a, b) \in I$ for all $a, b \in F$, $a \neq b$, are exactly the linearily ordered subsets of X. It follows that $\mu \in \mathbf{Z}\langle\langle X^*\rangle\rangle$ is a Möbius function for M. Let $\zeta \in \mathbf{Z}\langle\langle X^*\rangle\rangle$ be the formal inverse of μ. We show that $\zeta(w) = 1$ if $w \in X^*$ is irreducible and $\zeta(w) = 0$ else.

Clearly, we have $\mu(1) = \zeta(1) = 1$ and $\mu(a) = -1$, $\zeta(a) = 1$ for every letter $a \in X$.

Next, observe that every word $w \in X^*$, $w \neq 1$ has the form $w = a_1 \ldots a_m w'$ with $m \geq 1$, $a_i \in X$ for $1 \leq i \leq m$, $a_i < a_j$ for $1 \leq i < j \leq m$, $w' \in X^*$ and if $w' = a_0 w''$ for some $a_0 \in X$, $w'' \in X^*$ then we do not have $a_m < a_0$.

Lemma 4.3.4 *Let $w = a_1 \ldots a_m w'$ as above. If $m = 1$ then we have $\zeta(w) = \zeta(w')$ else we have $\zeta(w') = 0$.*

Proof of Lemma 4.3.4: For every $m \geq 1$ we have

$$-\zeta(w) = \sum_{k=1}^{m} \mu(a_1 \ldots a_k)\zeta(a_{k+1} \ldots a_m w')$$

Hence, for $m = 1$ we find $-\zeta(w) = \mu(a_1)\zeta(w') = -\zeta(w')$. If $m = 2$ then
$-\zeta(w) = \mu(a_1)\zeta(a_2 w') + \mu(a_1 a_2)\zeta(w') = -\zeta(a_2 w') + \zeta(w') = -\zeta(w') + \zeta(w') = 0$.
Let $m > 2$. Then by induction:

$$\begin{aligned} -\zeta(w) &= \mu(a_1 \ldots a_{m-1})\zeta(a_m w') + \mu(a_1 \ldots a_m)\zeta(w') \\ &= (-1)^{m-1}\zeta(w') + (-1)^m \zeta(w') = 0 \qquad \square \end{aligned}$$

Proof of Theorem 4.3.3 (continued):   Let $w = a_1 \ldots a_m w'$ as above. If $m \geq 2$ then w is reducible and $\zeta(w) = 0$ by Lemma 4.3.4. If $m = 1$ then w is irreducible if and only if w' is irreducible. Further, by Lemma 4.3.4 we have $\zeta(w) = \zeta(w')$ in this case. The result follows by induction since $\zeta(1) = 1$ and $1 \in \mathrm{Irr}(S)$. $\square$

Remark 4.3.5 In the proof of the theorem above we never used that $S = S(I^+)$ is complete. The completeness of S can be deduced from the theorem without inspecting any critical pairs: We have shown that $\mu^{-1} = \sum_{w \in \overline{\mathrm{Irr}}(S)} w$ for some Möbius function $\mu \in \mathbf{Z}\langle\langle X^* \rangle\rangle$. Hence $\mathrm{Irr}(S)$ must be in bijection with M. Since S is noetherian, this is enough to ensure its completeness. $\square$

If one does not want to involve semi-Thue systems the contents of Theorem 4.3.3 may be rephrased in the following way:

Corollary 4.3.6 *Let $<$ be any partial order on X. Let $L \subseteq X^*$ be the finite set of words*

$$L = \{a_1 \ldots a_m \mid m \geq 0, \ a_i \in X \text{ for } 1 \leq i \leq m, a_i < a_j \text{ for } 1 \leq i < j \leq m\}$$

and let $R \subseteq X^*$ be the complement of the rational set $\bigcup_{a<b} X^* ab X^*$. Define $\mu' = \sum_{w \in L}(-1)^{|w|}w$ and $\zeta' = \sum_{w \in R} w$. Then μ' and ζ' are formal inverses and the following assertions are equivalent:

i) μ' is a Möbius function;

ii) μ' is an unambiguous Möbius function;

iii) R is a rational cross-section of M in X^*.

Proof: Lemma 4.3.4 is valid for every partial order $<$ on X. It follows that μ' and ζ' are formal inverses. The implications i) $\Rightarrow$ ii), and ii) $\Rightarrow$ iii) are therefore consequences of Remark 4.3.2. To see iii) $\Rightarrow$ i), let R be a rational cross-section of M in X^*. The reverse operation, rev, on words which is inductively defined by $\mathrm{rev}(1) = 1$ and $\mathrm{rev}(aw) = \mathrm{rev}(w)a$ for $a \in X$, $w \in X^*$, is also defined for traces. Since rev maps the support of μ' injectively into R, it follows that the canonical projection restricted to $\mathrm{supp}(\mu')$ is injective. We have to show that $p(\mu') = \mu_M$ is the Möbius function in $Z\langle\langle M \rangle\rangle$ where $p : Z\langle\langle X^* \rangle\rangle \to Z\langle\langle M \rangle\rangle$ is the canonical ring-homomorphism. But this is purely formal by Proposition 4.3.1 and $1 = \zeta'\mu'$ and $p(\zeta') = \zeta_M = \sum_{w \in M} w \in Z\langle\langle M \rangle\rangle$. $\square$

The next corollary may be interpreted as a Möbius inversion formula:

Corollary 4.3.7 *(Same notations as in Corollary 4.3.6). Let $Q \subseteq X^*$ be any subset which is closed under suffixes, and let $f, g \in Z\langle\langle X^* \rangle\rangle$ be power series.*
     Then the following formulae are equivalent:

i) $\displaystyle\sum_{\substack{w_1 w_2 = w \\ w_1 \in R}} g(w_2) = f(w)$ *for all $w \in Q$,*

ii) $\displaystyle\sum_{\substack{w_1 w_2 = w \\ w_1 \in L}} (-1)^{|w_1|} f(w_2) = g(w)$ *for all $w \in Q$.*

Proof: i) $\Rightarrow$ ii): By i) we have $(\zeta'g)(w) = f(w)$ for every $w \in Q$ and the same holds for every suffix w_2 of $w \in Q$. Hence, for $w \in Q$ we obtain:

$$g(w) = (\mu'\zeta'g)(w) = \sum_{w_1 w_2 = w} \mu(w_1)(\zeta'g)(w_2)$$
$$= \sum_{w_1 w_2 = w} \mu(w_1)f(w_2)$$
$$= \sum_{\substack{w_1 w_2 = w \\ w_1 \in L}} (-1)^{|w_1|} f(w_2).$$

ii) $\Rightarrow$ i): analogous. $\square$

Remark 4.3.8   Since $R \cap L = \{1\} \cup X$, for each $w \in X^*$ the sum in formula i) or the sum in formula ii) has at most two terms. In the special case $Q = R$ we have the equivalence:

i) $\displaystyle\sum_{w_1 w_2 = w} g(w_2) = f(w)$ for all $w \in R$

ii) $f(1) = g(1)$ and,
     $f(aw) - f(w) = g(aw)$ for all $a \in X$, $w \in X^*$
     such that $aw \in R$. $\square$

Assume now we have given any unambiguous Möbius function $\mu \in Z\langle\langle X^* \rangle\rangle$. For $(a, b) \in I$ we either have $\mu(ab) = 1$, $\mu(ba) = 0$ or $\mu(ba) = 1$, $\mu(ab) = 0$. (This follows by the definition of an unambiguous Möbius function and $\mu_M([\{a, b\}]) = 1$.) Thus, the set $I^+(\mu) := \{(a, b) \in X \times X \mid \mu(ab) = 1\}$ is an orientation of I.

Theorem 4.3.9 *The orientation $I^+(\mu)$ is transitive.*

Proof: Let $\zeta \in \mathbf{Z}\langle\langle X^* \rangle\rangle$ be the formal inverse of μ. Set $I^+ = I^+(\mu)$. Let $(a, b), (b, c) \in I^+$. We have to show $(a, c) \in I^+$. Consider

$$
\begin{aligned}
-\zeta(abc) &= \mu(a)\zeta(bc) + \mu(ab)\zeta(c) + \mu(abc) \\
&= \mu(a)(-\mu(b)\zeta(c) - \mu(bc)) + \zeta(c) + \mu(abc) \\
&= -1(1-1) + 1 + \mu(abc) = 1 + \mu(abc)
\end{aligned}
$$

Since $\zeta(abc) \geq 0$ we have $\mu(abc) = -1$. In particular, we already have $(a, c) \in I$. Therefore the words abc and bca denote the same trace in M and $\mu(abc) = -1$ implies $\mu(bca) = 0$.

Thus,

$$
\begin{aligned}
-\zeta(bca) &= \mu(b)\zeta(ca) + \mu(bc)\zeta(a) \\
&= -\zeta(ca) + 1 = \mu(c)\zeta(a) + \mu(ca) + 1 \\
&= \mu(ca) \leq 0
\end{aligned}
$$

Hence, $\mu(ca) = 0$. Therefore $\mu(ac) = 1$; and this proves $(a, c) \in I^+$. $\square$

Corollary 4.3.10 *Let X be a finite alphabet, $I \subseteq X \times X$ be an independence relation, and $M = X^*/\{(ab, ba) \mid (a, b) \in I\}$. Then the following assertions are equivalent:*

i) There is a transitive orientation of I.

ii) There is an unambiguous Möbius function for M in $\mathbf{Z}\langle\langle X^* \rangle\rangle$.

Proof: Combine Theorem 4.3.3 and Theorem 4.3.9. $\square$

Example: i) Let (X, I) be given by the independence graph:

$$ a \quad\text{------}\quad b \quad\text{------}\quad c \quad\text{------}\quad d. $$

Then $a \leftarrow b \rightarrow c \leftarrow d$ is a transitive orientation and

$$ \mu = 1 - a - b - c - d + ba + bc + dc $$

is an unambiguous Möbius function.

More generally if X has at most four letters then for each independence relation there is always a transitive orientation and hence always an unambiguous Möbius function.

ii) Let (X, I) be given by the graph

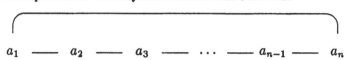

$$a_1 \quad\text{———}\quad a_2 \quad\text{———}\quad a_3 \quad\text{———}\quad \cdots \quad\text{———}\quad a_{n-1} \quad\text{———}\quad a_n$$

with $n \geq 4$. Then there exists a transitive orientation if and only if n is even. In particular there is no unambiguous Möbius function for the trace monoid which is associated with the independence graph

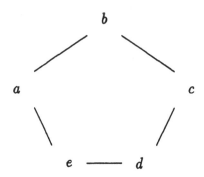

So far, we have associated to each transitive orientation an unambiguous Möbius function and to each unambiguous Möbius function a transitive orientation. The question is whether there are other unambiguous Möbius functions. This is not the case but surprisingly the proof is rather complicated compared to the other computations of this section.

Theorem 4.3.11 *Let X be a finite alphabet, $I \subseteq X \times X$ be an independence relation, and $M = X^*/\{(ab, ba) \mid (a, b) \in I\}$. Then the set of transitive orientations of I is in canonical one-to-one correspondence with the set of unambiguous Möbius functions for M in $\mathbf{Z}\langle\langle X^*\rangle\rangle$.*

Proof: It is clear from the construction that we have $I^+ = I^+(\mu(I^+))$ for every transitive orientation. Let μ be an unambiguous Möbius function and $I^+ = I^+(\mu)$ the associated transitive orientation. Define $\mu' := \mu(I^+) \in \mathbf{Z}\langle\langle X^*\rangle\rangle$, then μ' is an unambiguous Möbius function by Theorem 4.3.3. We have to show $\mu' = \mu$. From the construction it is clear that $\mu'(w) = \mu(w)$ for every word of length at most two. The problem is to show this equality for longer words. As usual, we write $a < b$ for $(a, b) \in I^+$, i.e., $\mu(ab) = 1$. Observe that it is enough to show $\mu'(a_1 \ldots a_n) = \mu(a_1 \ldots a_n)$ for $n > 2$, $a_i \in X$ and $a_1 < a_2 < \ldots < a_n$.

Let $\zeta \in \mathbf{Z}\langle\langle X^*\rangle\rangle$ denote the formal inverse of μ'. By definition it is a characteristic function of a set of representatives for M. Assume we would have $\mu' \neq \mu$, then there is a smallest integer $n \geq 2$ and a sequence of linearly ordered letters $a_1 < a_2 < \ldots < a_n$ such that $\mu'(a_1 \ldots a_n) \neq (-1)^n$.

Claim: Let $a_1 < \ldots < a_m$ be a sequence of linearly ordered letters with $1 \leq m < n$. Then we have $\zeta(a_m \ldots a_1) = 1$ and $\zeta(a_{\sigma(m)} \ldots a_{\sigma(1)}) = 0$ for every non-trivial permutation σ of $\{1, \ldots, m\}$.

Proof: To prove the claim, it is enough to show that $\zeta(a_m \ldots a_1) = 1$. For $m = 1$ this is obvious. For $m \geq 2$ the claim follows by induction since for $m < n$ we have $\mu'(a_1 \ldots a_m) = \mu(a_1 \ldots a_m)$, hence $-\zeta(a_m \ldots a_1) = -\zeta(a_{m-1} \ldots a_1)$. $\square$

Proof of 4.3.11 (continued):   From the claim we obtain

$$
\begin{aligned}
-\zeta(a_1 \ldots a_n) &= \mu'(a_1 \ldots a_{n-1})\zeta(a_n) + \mu'(a_1 \ldots a_n) \\
&= (-1)^{n-1} + \mu'(a_1 \ldots a_n)
\end{aligned}
$$

Since, by assumption, $\mu'(a_1 \ldots a_n) \neq \mu(a_1 \ldots a_n)$, we have $\mu'(a_1 \ldots a_n) = 0$ and n must be even. (Otherwise we would have $\zeta(a_1 \ldots a_n) < 0$.)

Let σ be the non-trivial permutation of $\{1, \ldots, n\}$ such that $\mu'(a_{\sigma(1)} \ldots a_{\sigma(n)}) = 1$. Write $b_i := a_{\sigma(i)}$ for $i = 1, \ldots, n$. We have

$$
-\zeta(b_1 \ldots b_n) = \left(\sum_{k=1}^{n-1} \mu(b_1 \ldots b_k)\zeta(b_{k+1} \ldots b_k) \right) + 1
$$

Thus, there is a smallest integer k, $1 \leq k \leq n-1$ such that

$$
\mu(b_1 \ldots b_k)\zeta(b_{k+1} \ldots b_k) \neq 0
$$

Let k be this integer. By the claim above we have $b_1 < \cdots < b_k$ and $b_{k+1} > \cdots > b_n$.

If we would have $b_k < b_{k+1}$ then $k < n-1$ since $b_1 \ldots b_n \neq a_1 \ldots a_n$. But then

$$
\begin{aligned}
-\zeta(b_1 \ldots b_n) &= \mu(b_1 \ldots b_k)\zeta(b_{k+1} \ldots b_n) \\
&\quad + \mu(b_1 \ldots b_{k+1})\zeta(b_{k+2} \ldots b_n) \\
&\quad + 1 \\
&= (-1)^k + (-1)^{k+1} + 1 = 1
\end{aligned}
$$

which is impossible.

Hence it holds $b_k > b_{k+1}$. If we would have $k > 1$ then we would have $\mu(b_1 \ldots b_{k-1}) \, \zeta(b_k \ldots b_n) \neq 0$ contradicting the minimality of k. It follows $k = 1$ and $b_1 > b_2 > \cdots > b_n$.

This means $b_1 \ldots b_n = a_n \ldots a_1$. We obtain $\mu'(a_n \ldots a_1) = 1$ and

$$
-\zeta(a_n \ldots a_1) = \mu(a_n)\zeta(a_{n-1} \ldots a_1) + \mu'(a_n \ldots a_1) = -1 + 1 = 0
$$

But this leads to the following contradiction:

$$
\begin{aligned}
0 \geq -\zeta(a_{n-1}a_n a_{n-1} \ldots a_1) &= \mu(a_{n-1})\zeta(a_n \ldots a_1) + \mu(a_{n-1}a_n)\zeta(a_{n-1} \ldots a_1) \\
&= 0 + 1 = 1.
\end{aligned}
$$

Thus $\mu'(a_1 \ldots a_n) \neq \mu(a_1 \ldots a_n)$ is impossible which concludes the proof of Theorem 4.3.11 □

From the correspondence between transitive orientations and unambiguous Möbius functions we may immediately deduce the following corollaries. Of course, these corollaries can be verified by a direct computation, too. They were independently also obtained by Roman König, (personal communication).

Corollary 4.3.12 *Let M be a trace monoid generated by some alphabet X, and let $M' \subseteq M$ be a submonoid generated by some subset $X' \subseteq X$.*
*   If $\mu \in \mathbf{Z}\langle\langle X^* \rangle\rangle$ is an unambiguous Möbius function for M then the restriction*
$$\mu' = \sum_{w \in X'^\bullet} \mu(w)w \in \mathbf{Z}\langle\langle X'^* \rangle\rangle \text{ is an unambiguous Möbius function for } M'. \ \square$$

Corollary 4.3.13 *Let M_i be a trace monoid generated by X_i such that M_i has an unambiguous Möbius function $\mu_i \in \mathbf{Z}\langle\langle X_i^* \rangle\rangle$, $i = 1, 2$. Let X denote the disjoint union of X_1 and X_2. Then we have:*

i) *The function*

$$\mu_1 + \mu_2 - 1 = 1 + \sum_{\substack{w_1 \in X_1^\bullet \\ w_1 \neq 1}} \mu(w_1)w_1 + \sum_{\substack{w_2 \in X_2^\bullet \\ w_2 \neq 1}} \mu(w_2)w_2 \in \mathbf{Z}\langle\langle X^* \rangle\rangle$$

is an unambiguous Möbius function for the free product $M_1 * M_2$.

ii) *The function*

$$\mu_1 \mu_2 = \sum_{\substack{w_1 \in X_1^\bullet \\ w_2 \in X_2^\bullet}} \mu(w_1)\mu(w_2)w_1 w_2 \in \mathbf{Z}\langle\langle X^* \rangle\rangle$$

is an unambiguous Möbius function for the direct product $M_1 \times M_2$. □

Remark 4.3.14 It follows by Corollary 4.3.13 that every direct product of free monoids has an unambiguous Möbius function. Since every trace monoid is a submonoid of such a direct product, we see that the existence of unambiguous Möbius functions does not transfer to submonoids, in general. (Compare this with the assertion from Corollary 4.3.12.)

Let us now return to the discussion of the existence of finite complete semi-Thue systems for over X defining M. Directly from Proposition 4.2.3 we obtain

Theorem 4.3.15 *Let X be a finite alphabet, $I \subseteq X \times X$ be an independence relation, and $M = X^*/\{(ab, ba) \mid (a, b) \in I\}$. Then the following three sets are in canonical one-to-one correspondence:*

i) transitive orientations of I;

ii) unambiguous Möbius functions for M in $\mathbf{Z}\langle\langle X^* \rangle\rangle$;

iii) finite normalized complete semi-Thue systems over X which define M. □

The question whether there is a transitive orientation of I is purely graph-theoretical. It asks whether the independence graph (X, I) is a so-called comparability graph. It is well-known that this question can be answered in polynomial time, see [Gol86, Chap. 5.6]. Thus analogously to [Ott87, Cor. to Thm. 2] we may state:

Corollary 4.3.16 *There is a polynomial time algorithm for the following problem:*

Given a finite alphabet X with independence relation $I \subseteq X \times X$.

Decide whether $M = X^*/\{(ab, ba) \mid (a, b) \in I\}$ has an unambiguous Möbius function in $\mathbf{Z}\langle\langle X^* \rangle\rangle$ and if it has then compute such a function. □

Remark 4.3.17 In a forthcoming paper we will generalize the results above to a relative situation $p : M_1 \to M_2$ where $M_j = X^*/\{(ab, ba) \mid (a, b) \in I_j\}$, $j = 1, 2$ with independence relations $I_1 \subseteq I_2$. (Note that we considered here the special case: $I_1 = \emptyset$.)

4.4 Möbius Functions and Semi-Thue Systems

Let $S = X^* \times X^*$ be any semi-Thue system. We are going to determine the formal inverse in $\mathbf{Z}\langle\langle X^* \rangle\rangle$ of the characteristic series over the irreducible words. Of course, this requires that 1 is irreducible, which we shall henceforth assume. For noetherian systems this requirement is always true. Our approach has some nice combinatorial applications. For example it yields, as a special case, a new proof of Lemma 4.3.4 above which need no real computation. Slightly more general, we start with any non-empty subset $R \subseteq X^*$ which is closed with respect to factors, for example $R = \mathrm{Irr}(S)$ for a semi-Thue system S. The complement $X^* \setminus R$ is an ideal which is generated by some basis $L \subseteq X^*$, i.e., $1 \in R = X^* \setminus X^*LX^*$ and $L \cap X^+LX^* = L \cap X^*LX^+ = \emptyset$. (For a system $S \subseteq X^+ \times X^*$ the set L is the set of left-hand sides where no proper factors are other left-hand sides.)

We need a definition. An *overlapping-chain with respect to L* is a sequence $(w_1, \ldots, w_n)$, $n \geq 0$ of words such that $w_1 \in X$, $w_i \in R$ for $1 \leq i \leq n$ and $w_i w_{i+1} \in X^*L \setminus X^*LX^+$ for $1 \leq i < n$. This means w_1 is a letter, no factor

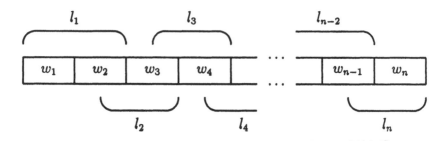

Figure 4.1. An overlapping chain

of w_i is in L, a (unique) suffix of $w_i w_{i+1}$ is in L and no other factor of $w_i w_{i+1}$ belongs to L. In particular, $w_i \neq 1$ for all $1 \leq i \leq n$. We have the following basic property:

Lemma 4.4.1 *Let* $(u_1, \ldots, u_m)$ *and* $(v_1, \ldots, v_n)$ *be overlapping-chains such that the word* $u_1 \ldots u_m$ *is a prefix of* $v_1 \ldots v_n$. *Then we have* $m \leq n$ *and* $u_i = v_i$ *for all* $1 \leq i \leq m$.

Proof: Let $0 \leq i \leq \min(m, n)$ be maximal such that $(u_1, \ldots, u_i) = (v_1, \ldots, v_i)$. If $i = \min(m, n)$ then obviously $i = m \leq n$ and the result follows. Hence assume $i < \min(m, n)$. Since u_1, v_1 are letters, we have $1 \leq i$. Further we have that u_{i+1} is a proper prefix of v_{i+1} or vice versa. Say, u_{i+1} is a proper prefix of v_{i+1}. Then $u_i u_{i+1}$ is a proper prefix of $v_i v_{i+1}$, too. But this implies $v_i v_{i+1} \in X^* L \cap X^* L X^+$, contradicting the definition of an overlapping-chain. $\square$

Let $\mathcal{L}$ be the set of overlapping-chains with respect to L. Then the mapping $\mathcal{L} \to X^*$, $(w_1, \ldots, w_n) \mapsto w_1 \ldots w_n$ is injective by the lemma above. Thus, we may identify $\mathcal{L}$ with a subset of X^*. If $w = (w_1, \ldots, w_n) \in \mathcal{L} \subseteq X^*$ is an overlapping-chain we have a picture as in Figure 4.1 where $l_1, \ldots, l_{n-1} \in L$ and no other factor of w belong to L. Short overlapping-chains have direct interpretations:

A letter of X is an overlapping-chain if and only if this letter is not in L, overlapping-chains of length two are the elements of $\{l \in L \mid |l| \geq 2\}$.

If $S = X^* \times X^*$ is a noetherian semi-Thue system where no left-hand side of a rule is factor of any other left-hand side, then the overlapping-chains of length three correspond exactly to the minimal critical pairs of S. Thus by Theorem 4.1.1 such a system is confluent if and only if it confluent on all overlapping-chains of length three. Therefore one is led to consider minimal critical pairs by looking for an interpretation of short overlapping-chains. Also the restriction of

Theorem 4.1.1 concerning the left-hand sides is very natural in this context: It simply says $L = \{l \mid (l, r) \in S\}$.

The reason why we are interested in overlapping-chains here, is the following identity of formal power series:

Theorem 4.4.2 *Let $L \subseteq X^*$ and $R \subseteq X^*$ as above, i.e., L is the basis of an ideal and $R = X^* \setminus X^* L X^*$, and let $\mathcal{L}$ be the set of overlapping-chains with respect to L. Define:*

$$\mu_{\mathcal{L}} = \sum_{(w_1, \ldots, w_n) \in \mathcal{L}} (-1)^n w_1 \ldots w_n$$
$$\zeta_R = \sum_{w \in R} w$$

Then the power series $\mu_{\mathcal{L}}, \zeta_R \in \mathbf{Z}\langle\langle X^* \rangle\rangle$ are formal inverses of each other.

Proof: Following ideas of [Vie86], we give a "bijective proof" by defining an involution without fixed points ψ on the set $\mathcal{L} \times R \setminus \{(1,1)\}$. Let $(w, r) \in \mathcal{L} \times R$, $(w, r) \neq (1, 1)$.
Define $\psi(w, r) = (w', r')$ as follows:
   If $(w, r) = (1, r_1 r')$ with $r_1 \in X$, $r' \in X^*$ then set $(w', r') = (r_1, r')$.
   If $(w, r) = ((w_1, \ldots, w_n), r)$ such that $w_n r \in R$
then set $(w', r') = ((w_1, \ldots, w_{n-1}), w_n r)$.
   If $(w, r) = ((w_1, \ldots, w_n), r)$ such that $w_n r \notin R$ then we obtain a factorization $w_n r = w_n w_{n+1} r'$ such that $(w_1, \ldots, w_n, w_{n+1})$ is an overlapping-chain and, by the lemma above, this factorization is unique. Set $(w', r') = ((w_1, \ldots, w_n, w_{n+1}), r')$ in this case.
   Obviously in all cases $\psi(w', r') = (w, r)$ and it holds: $\mu_{\mathcal{L}}(w)\zeta_R(r) = -\mu_{\mathcal{L}}(w')\zeta_R(r')$ and $wr = w'r'$ for all $(w, r) \in \mathcal{L} \times R \setminus \{(1, 1)\}$. The result follows since $\mu_{\mathcal{L}}\zeta_R = 1 + \sum_{\substack{(w,r) \in \mathcal{L} \times R \\ (w,r) \neq (1,1)}} \mu_{\mathcal{L}}(w)\zeta_R(r)wr.$ $\square$

Let us state some applications of the identy above:
1): Let $\rho \subseteq X \times X$ be any relation, $\bar{\rho} = X \times X \setminus \rho$ and $L = \{ab \mid (a, b) \in \bar{\rho}\}$. Then we have

$$R = \{a_1 \ldots a_n \mid n \geq 0, (a_i, a_{i+1}) \in \rho \text{ for } 1 \leq i \leq n\}$$

and

$$\mathcal{L} = \{a_1 \ldots a_m \mid m \geq 0, (a_i, a_{i+1}) \in \bar{\rho} \text{ for } 1 \leq i < n\} = \overline{R}.$$

Hence, $\mu_{\mathcal{L}}\zeta_R = \mu_{\overline{R}}\zeta_R = \mu_R \zeta_{\overline{R}} = 1 \in \mathbf{Z}\langle\langle X^* \rangle\rangle$.
In particular, if X is partially ordered and $\bar{\rho}$ is the relation $<$ then the identity yields Lemma 4.3.4.

2): Let $S \subseteq X^* \times X^*$ be a semi-Thue system with $1 \in \mathrm{Irr}(S)$, $L \subseteq \{l \in X^* \mid (l, r) \in S\}$ such that L is the basis of the ideal $X^* \setminus \mathrm{Irr}(S)$, and $\mathcal{L}$ let be the corresponding set of overlapping-chains. Set $\mu(S) = \mu_{\mathcal{L}}$. The function $\mu(S)$ tells us something about the number of factorizations of a word $x \in X^*$ into irreducible factors. Indeed, let $\mu^+(S)(w)$ ($\mu^-(S)(w)$ resp.) be the number of factorizations $w = w_1 \ldots w_n$ with $1 \neq w_i \in \mathrm{Irr}(S)$ and n even (n odd resp.). Then it holds $\mu(S) = \mu^+(S) - \mu^-(S)$. This is clear, since the identity $(\mu^+(S) - \mu^-(S))\zeta_R = 1 \in \mathbf{Z}\langle\langle X^* \rangle\rangle$ is trivial for $R = \mathrm{Irr}(S)$.

3): For $k \geq 0$ and $S \subseteq X^* \times X^*$ as above let us define the following integers:

$$\begin{aligned}
\mu^+(k) &= \#\{(w_1, \ldots, w_n) \in \mathcal{L} \mid n \text{ even}, |w_1 \ldots w_n| = k\}, \\
\mu^-(k) &= \#\{(w_1, \ldots, w_n) \in \mathcal{L} \mid n \text{ odd}, |w_1 \ldots w_n| = k\}, \\
\zeta(k) &= \#\{w \in \mathrm{Irr}(S) \mid |w| = k\}.
\end{aligned}$$

Then we obtain the following identity of formal power series in one variable:

$$\sum_{k \geq 0} \zeta(k) z^k = \frac{1}{\left(\sum_{k \geq 0} (\mu^+(k) - \mu^-(k)) z^k \right)} \qquad \square$$

Finally, let us consider a noetherian semi-Thue system $S \subseteq X^* \times X^*$ such that each congruence class modulo S of X^* is finite. Then the number of factorizations $w = w_1 w_2$ modulo S is finite for all $w \in X^*$.

Hence, the monoid X^*/S has a Möbius-function in the sense of Cartier Foata, see [CF69]; and the canonical projection induces a surjective ring homomorphism

$$p : \mathbf{Z}\langle\langle X^* \rangle\rangle \to \mathbf{Z}\langle\langle X^*/S \rangle\rangle$$

The theorem above now relates the completeness of S to the Möbius function of the quotient monoid:

Corollary 4.4.3 *Let $S = X^* \times X^*$ be a noetherian semi-Thue system such that the congruence classes $\{v \in X^* \mid v \overset{*}{\underset{S}{\Longleftrightarrow}} w\}$ are finite for all $w \in X^*$. Then the following assertions are equivalent:*

i) The system S is complete.

ii) The power series $p(\mu(S)) \in \mathbf{Z}\langle\langle X^*/S \rangle\rangle$ is the Möbius function of the monoid X^*/S.

Proof: The result follows since the noetherian system S is complete if and only if it holds $p(\sum_{w \in \mathrm{Irr}(S)} w) = \zeta_M \in \mathbf{Z}\langle\langle X^*/S \rangle\rangle$. $\square$

The main application of the above corollary lies in the direction i) $\Rightarrow$ ii). It is usually a very difficult task to compute Möbius functions of monoids. Now, if we can present a monoid as X^*/S with a complete system S as above then we have found a description of the Möbius function in terms of overlapping-chains.

Example: Let $X = \{a, b, c\}$ and $S = \{abc \Longrightarrow cb\}$. Then S is a complete semi-Thue system such that all congruence classes modulo S are finite. (We have considered the monoid $M = X^*/S$ earlier as an example where the divisor relation is noetherian but no weight function exists.)

Then $1 - a - b - c + abc \in \mathbf{Z}\langle\langle M \rangle\rangle$ is the Möbius function of M. $\square$

5. Trace Replacement Systems

5.1 Preliminaries

In order to define replacement systems over traces it is convenient to develop the relevant notions in a more general framework. These notions are straightforward from the semi-Thue case of Chap. 4.

Let M be any monoid with neutral element 1. A *replacement system* over M is a subset $S \subseteq M \times M$. It defines a relation $\underset{S}{\Longrightarrow}$ by $ulu' \underset{S}{\Longrightarrow} uru'$ for $u, u' \in M$, $(l, r) \in S$. The elements of S are called *rules*, again often written in the form $l \Longrightarrow r$ for $(l, r) \in S$. A replacement system over a free monoid is therefore a semi-Thue system, and a replacement system over $\mathbb{N}^k$, $k \geq 1$, is called a *vector replacement system*. If M is a trace monoid then we shall speak of a *trace replacement system*. As usual, $\underset{S}{\overset{+}{\Longrightarrow}}$, $\underset{S}{\overset{*}{\Longrightarrow}}$ and $\underset{S}{\overset{*}{\Longleftrightarrow}}$ denote the transitive closure, the reflexive, transitive closure and the symmetric, reflexive, transitive closure of $\underset{S}{\Longrightarrow}$. The relation $\underset{S}{\overset{*}{\Longleftrightarrow}}$ is a congruence, and we write M/S for its quotient monoid. Frequently bars are used to denote images of M in M/S. By $\mathrm{Irr}(S)$ we understand the set of irreducible elements in M, i.e., the set of elements where no rule of S applies to. The complement of $\mathrm{Irr}(S)$ is denoted by $\mathrm{Red}(S)$. This is the ideal $\mathrm{Red}(S) = \bigcup_{(l,r) \in S} MlM$. Note that if S is finite then $\mathrm{Red}(S)$ is a rational set. A system $S \subseteq M \times M$ is called *normalized* if for every rule $(l, r) \in S$ the right-hand side r is irreducible and the left-hand side l is irreducible with respect to $S \setminus \{(l, r)\}$. Two systems $S \subseteq M \times M$ and $T \subseteq M \times M$ are called *equivalent* if $M/S = M/T$. A system $S \subseteq M \times M$ is called *noetherian* if there are no infinite chains of the form $u_0 \underset{S}{\Longrightarrow} u_1 \underset{S}{\Longrightarrow} u_2 \underset{S}{\Longrightarrow} u_3 \ldots$. It is called *confluent* if for all $u, v, w \in M$ with $u \underset{S}{\overset{*}{\Longleftarrow}} v \underset{S}{\overset{*}{\Longrightarrow}} w$ there is a $z \in M$ such that $u \underset{S}{\overset{*}{\Longrightarrow}} z \underset{S}{\overset{*}{\Longleftarrow}} w$. A noetherian system $S \subseteq M \times M$ is confluent if and only if it is *locally confluent*, i.e. for all $u, v, w \in M$ with $u \underset{S}{\Longleftarrow} v \underset{S}{\Longrightarrow} w$ there is a $z \in M$ such that $u \underset{S}{\overset{*}{\Longleftrightarrow}} z \underset{S}{\overset{*}{\Longleftarrow}} w$. We say that a system is *complete* if it is noetherian and (locally) confluent. For complete systems the set $\mathrm{Irr}(S)$ is in canonical bijection with the quotient monoid M/S; thus complete systems provide us with normal forms in M for the elements of the quotient. In the following we make the technical assumption on all our systems that 1 never occurs on the left-hand side. Noetherian systems fulfill this assumption trivially since otherwise every element would be reducible.

Often the following situation occurs: We start with some system $T \subseteq M \times M$, and we are looking for a complete system $S \subseteq M \times M$ which is equivalent to T. In order to prove that S is such a system the following characterization is useful.

Proposition 5.1.1 *Let $S, T \subseteq M \times M$ be replacement systems such that S is noetherian and $S \subseteq \underset{T}{\overset{*}{\Longleftrightarrow}}$. Then the following assertions are equivalent:*

i) S is confluent and equivalent to T.

ii) The canonical mapping $M \to M/T$ restricted to $\mathrm{Irr}(S)$ is injective.

Proof: i) $\Rightarrow$ ii): trivial. ii) $\Rightarrow$ i): Since $S \subseteq \underset{T}{\overset{*}{\Longleftrightarrow}}$ we have the following sequence of canonical mappings:

$$\mathrm{Irr}(S) \overset{f}{\longrightarrow} M/S \overset{g}{\longrightarrow} M/T.$$

Since S is noetherian, f is surjective. Since gf is injective by ii), f is injective too, hence f is bijective. In particular, S is confluent. Now, since gf is injective and f is bijective, g is also injective. Since the surjectivity of g is clear, the mapping g is bijective, hence the equivalence of S and T. $\square$

Remark 5.1.2 If $S \subseteq M \times M$ is not noetherian then it may happen that $\mathrm{Irr}(S) \to M/T$ is bijective but S is neither confluent nor equivalent to T. To see this consider the following semi-Thue systems:

$$\begin{aligned} S &= \{a \Rightarrow b, a \Rightarrow a^2, b \Rightarrow b^2\} \\ T &= \{a \Rightarrow b, b \Rightarrow 1\}. \end{aligned}$$

We have $\{1\} = \mathrm{Irr}(S) = \{a, b\}^*/T$ but neither S is locally confluent nor $\{a, b\}^*/S$ is the trivial monoid. $\square$

An *admissible well-ordering* on M is a well-ordering $\prec$ of M such that $x \prec y$ implies $uxu' \prec uyu'$ for all $u, u', x, y \in M$. (We use linear orderings $\prec$ only; the generalization to partial orders is straightforward. It is not done here for simplification of the formalism.) If M has such an ordering then $uxu' = uyu'$ implies $x = y$, hence M is cancellative. Furthermore, M is group-free, i.e., $uv = vu = 1$ implies $u = v = 1$, and the divisor relation of M is noetherian, i.e., if $uxu' = y$ with $uu' \neq 1$ then $x \prec y$. It follows that every ideal I of M, (i.e. every subset $\emptyset \neq I \subseteq M$ of the form $I = MIM$), contains a set which is minimal in I for the proper divisor relation. This set is called a basis of I and is uniquely determined. Admissible well-orderings are used for replacement systems to ensure the noetherian condition. If M has an admissible well-ordering then every replacement system is equivalent to a noetherian one. Even more, we have the following fact:

Proposition 5.1.3 *Let M be a monoid with an admissible well-ordering. Then every quotient monoid of M is isomorphic to some M/S where $S \subseteq M \times M$ is a complete normalized replacement system.*

Proof: Let $p : M \to N$ be a surjective homomorphism. Since M has an admissible well-ordering $\prec$, for each $m \in M$ there is a unique $\hat{m} \in M$ which is minimal with respect to $\prec$ such that $p(m) = p(\hat{m})$. The set $I = \{m \in M \mid m \neq \hat{m}\}$ is an ideal since $\hat{m} \prec m$ implies $u\hat{m}v \prec umv$. Hence I is generated by some basis $L \subseteq I$. Furthermore, the quotient N is given by M/T where $T = \{(m, \hat{m}) \mid m \in I\}$. Define $S = \{(l, \hat{l}) \mid l \in L\}$, then S is noetherian and $S \subseteq \stackrel{*}{\underset{T}{\Longleftrightarrow}}$. Since $\mathrm{Irr}(S) = \{\hat{m} \mid m \in M\}$ is obviously in bijection with the quotient monoid N, it follows by Proposition 5.1.1 that p induces an isomorphism between M/S and N and that S is complete. Note that S is normalized by construction. $\Box$

Particularly useful are admissible well-orderings of *Knuth-Bendix type*. These are orderings such that for some weight function $\gamma : M \to \mathbb{N}$, (i.e. a homomorphism from M to $\mathbb{N}$ with $\gamma^{-1}(0) = \{1\}$), the inequality of weights $\gamma(x) < \gamma(y)$ implies $x \prec y$ for all $x, y \in M$. Typical examples of admissible well-ordering of Knuth-Bendix type are the lexical orderings on free monoids X^* or on free commutative monoids $\mathbb{N}^k$, $k \geq 1$. For $\mathbb{N}^k$, $k \geq 0$, it is well-known that all ideals are finitely generated, this is Dickson's Lemma [Dic13]. Since the complete system $S \subseteq M \times M$ mentioned in the proposition above is finite if and only if the ideal $I = \{\hat{m} \mid m \in \mathbb{N}\}$ has a finite basis, we obtain, as a special case, the following

Corollary 5.1.4 *i) [Red63], Every finitely generated abelian monoid is finitely presentable.*
ii) [Buc70], Every finitely generated abelian monoid is presentable by some finite normalized complete vector replacement system. $\Box$

Remark 5.1.5 The assertion of 5.1.4 ii) has been shown later in [Gil71] and [BL81], too. In [Die86a] it is shown that finitely generated abelian monoids are also presentable by some finite complete semi-Thue systems over free (non-commutative) monoids. But as in the case of trace monoids we have to add new generators, in general. $\Box$

The existence of an admissible well-ordering (of Knuth-Bendix type) transfers to submonoids and is invariant under isomorphisms. Since it is a non-trivial property, this implies that it is a so-called Markov property. It is a well-known classical result that Markov properties are undecidable, [Mar47]. Hence in general, it is undecidable whether a given finitely presented monoid M has such an ordering. However, we have the following positive result.

Proposition 5.1.6 *Every submonoid of a direct product of finitely many monoids with admissible well-orderings of Knuth-Bendix type has such an ordering, too.*

Proof: For $i = 1, 2$ let M_i be a monoid with an admissible well ordering $\prec_i$ of Knuth-Bendix type for the weight function $\gamma_i : M_i \to \mathbb{N}$. Then $\gamma : M_i \times M_2 \to \mathbb{N}$, $\gamma(m_1, m_2) := \gamma_1(m_1) + \gamma_2(m_2)$ is a weight function. Define the ordering $\prec$ by $(m_1, m_2) \prec (n_1, n_2)$ if $\gamma(m_1, m_2) \prec \gamma(n_1, n_2)$ or $\gamma(m_1, m_2) = \gamma(n_1, n_2)$ and $(m_1 \prec_1 n_1$ or $m_1 = n_1$ and $m_2 \prec_2 n_2)$. $\square$

Corollary 5.1.7 *Every trace monoid has an admissible well-ordering of Knuth-Bendix type.*

Proof: This follows by the embedding theorem, see Corollary 1.4.7 $\square$

Sometimes, other well-orderings are more suitable. This increases the flexibility. We shall also use the following:

Proposition 5.1.8 *Every submonoid of a free product of monoids with admissible well-orderings of Knuth-Bendix type has such an ordering, too.*

Proof: Let $M = \underset{i \in J}{*} M_i$. The proof is based on the fact that the free monoid J^* over the index set J has a lexical ordering, (axiom of choice). However, to be more explicit we consider the case $J = \{1, 2\}$ only. The general case is analogous to the finite case of two monoids. For $i = 1, 2$ let M_i, $\prec_i$, $\gamma_i : M_i \to \mathbb{N}$ as in the proof of the preceeding proposition. The weight functions γ_1, γ_2 induce a homomorphism $\gamma_{12} : M_1 * M_2 \to \{a, b\}^*$ by the extension of $\gamma_{12}(m_1) = a^{\gamma_1(m_1)}$ and $\gamma_{12}(m_2) = b^{\gamma_2(m_2)}$ for $m_1 \in M_1, m_2 \in M_2$. We use the lexical well-ordering $\prec_{12}$ on $\{a, b\}^*$ to define an ordering $\prec$ on $M_1 * M_2$ as follows: Let $m, m' \in M_1 * M_2, m \neq m'$. If $\gamma_{12}(m) \prec_{12} \gamma_{12}(m')$ then we set $m \prec m'$. If $\gamma_{12}(m) = \gamma_{12}(m')$ then we have $m = n_1 \ldots n_k$, $m' = n_1' \ldots n_k'$ with $k \geq 1$ and for each $j \in \{1, \ldots, k\}$ exists $i \in \{1, 2\}$ such that $n_j, n_j' \in M_i \setminus \{1\}$ and $n_{j-1} \in M_i \Rightarrow n_j \notin M_i$. Since $m \neq m'$ there is a minimal index $j \in \{1, \ldots, k\}$ such that $n_j \neq n_j'$. Define in this case $m \prec m'$ if $n_j \prec_i n_j'$. This is a well ordering of Knuth-Bendix type for the weight function $\gamma : M_1 * M_2 \to \mathbb{N}$, $\gamma(m) := |\gamma_{12}(m)|$ where $||$ denotes the length of words in $\{a, b\}^*$. $\square$

Warning: Let M be a trace monoid and assume that the underlying alphabet is totally ordered. Then we obtain a well-ordering on M if we compare traces by the lexical ordering of their lexicographic normal forms. But this is not an admissible well-ordering. Therefore it may not be used to show that trace replacement systems are noetherian. $\square$

To each noetherian replacement system $S \subseteq M \times M$ we canonically associate a partial ordering $\preceq_S$ of M as follows: We define $x \preceq_s y$ if $y \overset{*}{\underset{S}{\Longrightarrow}} uxv$ for some $u, v \in M$. As usual, $x \prec_S y$ means $x \preceq_s y$ and $x \neq y$. This partial ordering becomes important below because of its following basic property. Its proof is easy and left to the reader.

Lemma 5.1.9 *Let $S \subseteq M \times M$ be a noetherian replacement system and $\preceq_s$ be the canonically associated partial ordering above. If the proper divisor relation of M is noetherian then the ordering $\prec_S$ is well-founded, i.e., every non-empty subset of M has minimal elements with respect to $\prec_S$.* $\square$

Remark 5.1.10 Let $\preceq$ be an admissible well-ordering of M and $S \subseteq M \times M$ be a replacement system such that $l \succ r$ for all $(l, r) \in S$. Then the hypothesis of the lemma above is satisfied and $x \preceq_S y$ implies $x \preceq y$ for all $x, y \in M$. $\square$

5.2 The Knuth-Bendix Completion

The assertion of Proposition 5.1.3 is purely existential. It gives no answer how to find the complete system S nor whether S is finite. If S is finite then the Knuth-Bendix completion procedure is a general method to construct it. Before we recall the completion procedure, we have to define critical pairs for replacement systems over monoids. This is crucial for the following. Therefore, let us first informally explain the idea of critical pairs:

In order to test local confluence of a given system one usually starts with two rules p, q and considers all pairs resulting from elements of the monoid by application of these two rules. Call these pairs $\{p, q\}$-derivable. A subset of these $\{p, q\}$-derivable pairs is *critical* if in some sense it determines the confluence of all $\{p, q\}$-derivable pairs. This viewpoint is in the spirit of [Buc85]. It is formalized for our purposes where we work in monoids as follows.

We define critical pairs for noetherian systems only and in the definition we use the partial ordering $\preceq_S$ defined at the end of the last section.

Definition: Let $S \subseteq M \times M$ be a noetherian replacement system. A pair $(t_1, t_2) \in M \times M$ is called S-derivable if for some $t \in M$ we have $t_1 \underset{S}{\Longleftarrow} t \underset{S}{\Longrightarrow} t_2$.

A subset of S-derivable pairs is called critical if it contains (t_1, t_2) or (t_2, t_1) for all (t_1, t_2) such that the following holds:

For some $t \in M$ we have $t_1 \underset{S}{\Longleftarrow} t \underset{S}{\Longrightarrow} t_2$, the pair (t_1, t_2) is not confluent, but S is confluent on set of all $t' \in M$ where $t' \prec_S t$.

Any set of critical pairs is denoted by $CritPair(S)$.

To explain this definition consider the subset $N \subseteq M$ where S is not conflu-ent. If N is not empty then define N' to be the set of minimal elements of N

with respect to the ordering $\prec_S$, otherwise define $N' = \emptyset$. Then any subset of S-derivable pairs is denoted by $CritPair(S)$ if it satisfies

$$\{(t_1, t_2) \mid \exists t \in N' : t_1 \underset{S}{\Longleftarrow} t \underset{S}{\Longrightarrow} t_2, \quad (t_1, t_2) \text{ is not confluent}\}$$
$$\subseteq CritPair(S) \cup CritPair(S)^{-1}$$

Obviously, a noetherian system $S \subseteq M \times M$ is confluent if and only if all pairs of some set $CritPair(S)$ are confluent. More precisely, we have the following remark.

Remark 5.2.1 i): A noetherian system S has an empty set of critical pairs if and only if S is confluent.

ii): The set of all S-derivable pairs form a set of cirtical pairs.

iii): The set

$$\bigcup_{p,q \in S} CritPair(\{p, q\})$$

is a set of critical pairs for the whole system S. This allows to compute critical pairs locally on at most two rules.

iv): If S is a semi-Thue system or a vector replacement system then the usual set of critical pairs (as well as the minimal critical pairs as defined in Sect. 4.1) form a set of critical pairs $CritPair(S)$.

v): The definition of critical pairs given here differs from the definition given in [Die87a]. However, if a system $T \subseteq M \times M$ with $S \subseteq T$ is confluent on a set of critical pairs as defined in [Die87a], then it is also confluent on $CritPair(S)$ as defined above. In fact, it turned out that the former definition was slightly too strong to obtain good finiteness results. $\square$

In order to describe the Knuth-Bendix completion we assume that the monoid M has an admissible well-ordering $\prec$ such the set $\{x \in M \mid x \prec y\}$ is finite for all $y \in M$. We start with a replacement system $S_1 \subseteq M \times M$ such that $l \succ r$ for all $(l, r) \in S_1$. For $i \geq 1$ we define a system S_{i+1} inductively as follows. We pick any set $C(S_i)$ of S_i-derivable pairs. The only assumption on $C(S_i)$ is that if S_i is not confluent then $C(S_i)$ must contain at least one pair (t_1, t_2) which is not confluent and it holds $t_1 \underset{S_i}{\Longleftarrow} t \underset{S_i}{\Longrightarrow} t_2$ for some $t \in M$ such that S_i is confluent on all $t' \in M$ with $t' \prec t$. Thus, it is not necessary to consider all critical pairs of S_i in this step, but, of course, $C(S_i) = CritPair(S_i)$ will do. (The point is that all sets $CritPair(S)$ may be infinite whereas it is in principle enough to consider one pair after the other.) Define $S_{i+1} = S_i \cup \{(\tilde{t}_1, \tilde{t}_2) \mid \exists (t_1, t_2) \in C(S_i) : t_j \overset{*}{\Longrightarrow} \tilde{t}_j \in Irr(S_i), \ j = 1, 2, \{\tilde{t}_1, \tilde{t}_2\} = \{\hat{t}_1, \hat{t}_2\}, \hat{t}_1 \succ \hat{t}_2\}$. Finally, set $S^* = \bigcup_{i \geq 1} S_i$.

Proposition 5.2.2 i): *The system S^* defined above is complete and the normalization of S^* is exactly the system mentioned in Proposition 5.1.3.*

ii): Let $i \geq 1$. Then we have $S^* = S_i$ if and only if $S_i = S_{i+1}$.

iii): If there exists any finite complete system $T \subseteq M \times M$ such that $M/S_1 = M/T$ and $l \succ r$ for all $(l, r) \in T$ then we have $S^* = S_i = S_{i+1}$ for some $i \geq 1$.

iv): If S_1 is finite then there exists a sequence $S_1 \subseteq S_2 \ldots$ such that all $S_i \subseteq M \times M$ are finite. More precisely there exists such a sequence that for each $i \geq 1$ either S_{i+1} has exactly one rule more than S_i or $S^* = S_i$ and S_i is complete.

Proof: i): Assume that S^* would not be complete. Then there exists some $t_1 \underset{S^*}{\Longleftarrow} t \underset{S^*}{\Longrightarrow} t_2$ such that (t_1, t_2) is not confluent but S^* is confluent for all $t'_1 \underset{S^*}{\Longleftarrow} t' \underset{S^*}{\Longrightarrow} t'_2$ where $t' \prec t$. By assumption on the well-ordering $\prec$ for a given (t_1, t_2) there are only finitely many such pairs (t'_1, t'_2). Hence for some $i \geq 1$ we have $t_1 \underset{S_i}{\Longleftarrow} t \underset{S_i}{\Longrightarrow} t_2$ and all pairs (t'_1, t'_2) as above are confluent by S_i. Next consider the set $\{(t''_1, t''_2) \mid t''_1 \underset{S_i}{\Longleftarrow} t \underset{S_i}{\Longrightarrow} t''_2\}$ and denote it for a moment by C_i. The set C_i is finite, say of cardinality $k \geq 1$ and it contains (t_1, t_2). In $C(S_i)$ we have to pick at least one element of C_i, in $C(S_{i+1})$ another one and so forth. Indeed, let $(t''_1, t''_2) \in C(S_i)$, $t''_1 \underset{S_i}{\Longleftarrow} t'' \underset{S_i}{\Longrightarrow} t''_2$, such that (t''_1, t''_2) is not confluent but S_i is confluent on all t' where $t' \prec t''$. Then $t'' \preceq t$ since (t_1, t_2) is not confluent. But $t'' \prec t$ is impossible because otherwise (t''_1, t''_2) were of the type (t'_1, t'_2) as above. Hence $t'' = t$ and $(t''_1, t''_2) \in C_i$.

Now the completion procedure implies that (t_1, t_2) is confluent by S_{i+k}. This is a contradiction. Therefore S^* is complete. It is clear that the normal forms are the minimal representants of M/S_1 in M with respect to the well-ordering $\prec$. This implies that the normalization of S^* is the system constructed in Proposition 5.1.3.

ii): obvious.

iii): Let $T \subseteq M \times M$ be any finite system such that $l \succ r$ for all $(l, r) \in T$ and $M/S_1 = M/T$. Then for some $i \geq 0$ we have $T \subseteq \underset{S_i}{\overset{*}{\Longrightarrow}}$. If T is complete then S_i is complete, too. Hence $S^* = S_i = S_{i+1}$.

iv): This is clear since it is enough to pick $C(S_i)$ for $i \geq 1$ in such a way that contains at most one pair. $\square$

Remark 5.2.3 The description above of the Knuth Bendix completion transforms directly into a procedure if the following conditions are satisfied:

i) For all $x, y \in M$ we may test whether $x \prec y$.

ii) For all $y \in M$ the finite set $\{x \in M \mid x \prec y\}$ is computable.

iii) For all finite systems $S \subseteq M \times M$ a set of critical pairs $CritPair(S)$ is effectively constructible and $CritPair(S)$ is finite.

Proof: By i) the monoid has a decidable word problem. Together with ii) it allows us to test for given $t, l \in M$ whether $t = ulu'$ for some $u, u' \in M$. Thus, we may compute effectively irreducible descendants with respect to finite noetherian systems. By iii), S_{i+1} is constructible from S_i and it stays finite if S_i is finite. $\square$

Remark 5.2.4 Let M be finitely generated. If $\prec$ is an admissible well-ordering of Knuth-Bendix type. Then $\{x \in M \mid x \prec y\}$ is finite for all $y \in M$. If the weight is known for a set of generators then condition ii) above follows from i).

Proof: Let $\gamma : M \to \mathbb{N}$ be a weight function such that $\gamma(x) < \gamma(y)$ implies $x \prec y$. Now, for every $y \in M$ it holds $\{x \in M \mid x \prec y\} \subseteq \{x \in M \mid \gamma(x) \le \gamma(y)\}$. Since M is finitely generated the latter set is finite and computable. $\square$

5.3 Critical Pairs over Traces

By Proposition 5.1.3 and Corollary 5.1.7 every quotient monoid of a trace monoid has a presentation by some complete replacement system.

In order to use the Knuth-Bendix completion procedure for finding this complete system, we have to consider critical pairs. Our first result on critical pairs is negative. In general there are infinitely many critical pairs.

Theorem 5.3.1 *Let M be a free partially commutative monoid. If M is neither free nor commutative then there is a finite trace replacement system $S \subseteq M \times M$ where no finite set of critical pairs $CritPair(S)$ exists.*

Proof: Let $M = M(X, D)$. Since M is neither free nor commutative there are letters $a, b, c \in X$ such that $(a, b) \notin D$ and $(b, c) \in D$. Consider the noetherian system $S = \{ac \Rightarrow cc, \quad ca \Rightarrow cc\}$. If a and c are independent then S is a one-rule-system. However, in any case for all $k \ge 1$ we have $ccb^k c \underset{S}{\Longleftarrow} cab^k c = cb^k ac \underset{S}{\Longrightarrow} cb^k cc$ and $ccb^k c \ne cb^k cc$. We show that $CritPair(S)$ must contain these pairs $(ccb^k c, cb^k cc)$ for all $k \ge 1$. Indeed, let $t' \prec_S t = cab^k c$ for some $k \ge 1$. If t' is a factor of t then it contains at most one c or no a, and S is confluent on t'. If t' is a factor a descendant of t then t' contains no a, and S is confluent on t', too. According to the definition we have $(ccb^k c, cb^k cc) \in CritPair(S)$. $\square$

One might think that we have chosen the "wrong" definition of critical pairs and a "better" definition would yield finite sets. However, there can be no reasonable definition of critical pairs such that finite noetherian trace replacement systems have a finite and computable set of critical pairs. This is immediate from the following undecidability result proved by P. Narendran and F. Otto. We refer its proof to the original paper [NO88].

Note that the terminology of [NO88] is different. Instead of defining replacement systems over trace monoids they view them as certain so-called preperfect semi-Thue systems. But this approach is limited to consider only length-reducing systems.

Theorem 5.3.2 (Narendran/Otto) *There exists effectively a trace monoid M with only one pair of independent letters such that the confluence of finite length-reducing trace replacement systems over M is recursively undecidable.* $\square$

Remark 5.3.3 The minimal number of generators which are necessary to obtain the undecidability result above is unknown. For example, it is open whether we can decide confluence if M is generated by three letters only. It is also open whether we can decide confluence for finite monadic systems, i.e., for finite systems S where $|l| \geq 2 > |r|$ for all $(l, r) \in S$. $\square$

For our further investigation of critical pairs we need some notions and combinatorics on traces developped in the first chapter. Recall in particular that a subset $l \subseteq t$ is a subtrace if and only if $t = ulu'$ for some partition $t = u\dot\cup l\dot\cup u'$. We have shown in Lemma 1.2.5 that $l \subseteq t$ is a subtrace if and only if $x \leq y \leq z$ with $x, z \in l$ implies $y \in l$. For Hasse diagrams this is a global condition and not testable in the neighbourhood of l. (One should have this in mind when one thinks in terms of graph-grammars for replacements of traces.) If $l_1 \subseteq t$, $l_2 \subseteq t$ are subtraces then their intersection $l_1 \cap l_2$ and union $l_1 \cup l_2$ is defined with respect to t only. Note that $l_1 \cap l_2$ will be a subtrace whereas $l_1 \cup l_2$ may be viewed as a labelled subset only. Nevertheless, $l_1 \cup l_2$ defines a unique trace in M; but this will be, in general, no subtrace of t. The subtraces of elements *before*, *behind*, and *independent* of a given subtrace l have been defined by:

$$\begin{aligned}
\operatorname{pre}(l) &:= \{x \in t \setminus l \mid x \leq y \text{ for some } y \in l\}, \\
\operatorname{suf}(l) &:= \{x \in t \setminus l \mid y \leq x \text{ for some } y \in l\}, \\
\operatorname{ind}(l) &:= \{x \in t \setminus l \mid x \notin (\operatorname{pre}(l) \cup \operatorname{suf}(l))\}.
\end{aligned}$$

In particular, t is the disjoint union $t = \operatorname{pre}(l)\dot\cup l\dot\cup\operatorname{ind}(l)\dot\cup\operatorname{suf}(l)$. Further, if $t = ulu'$ then for some $v, v' \in M$ it holds: $u = \operatorname{pre}(l)v$, $vv' = \operatorname{ind}(l)$, and $v'\operatorname{suf}(l) = u'$.

Let $l \subseteq t$ by any subset of a trace $t \in M$. Then the subtrace of t generated by l has been defined by $\{y \in t \mid \exists x, z \in l : x \leq y \leq z\}$. We have denoted this subtrace by $\langle l \rangle$, it is the smallest subtrace of t containing l.

The following definitions are new:

We say that two subtraces $l_1 \subseteq t$, $l_2 \subseteq t$ are *strictly separated* if we have $l_i \subseteq (\operatorname{pre}(l_j) \cup \operatorname{ind}(l_j))$ for some $\{i, j\} = \{1, 2\}$. Hence $l_1 \subseteq t$, $l_2 \subseteq t$ are strictly separated if and only if for some $\{i,j\} = \{1,2\}$ we can write $t = ul_ivl_jw$ with $u = \mathrm{pre}(l_i)$, $w = \mathrm{suf}(l_j)$.

We say that $x \in t\backslash(l_1 \cup l_2)$ is *between* l_1 and l_2 if $\mathrm{pre}(x) \cap l_1 \neq \emptyset$, $\mathrm{suf}(x) \cap l_2 \neq \emptyset$ or $\mathrm{suf}(x) \cap l_1 \neq \emptyset$, $\mathrm{pre}(x) \cap l_2 \neq \emptyset$. Note that if x is between l_1 and l_2 then x must be independent of the intersection $l_1 \cap l_2$ and x belongs to the generated subtrace $\langle l_1 \cup l_2 \rangle$.

Example: We represent traces by their Hasse diagrams.
1) Let t be the following trace:

Then the subtraces $l_1 = (a \to b)$ and $l_2 = (c \to d)$ have an empty intersection (i.e., no overlapping) but they are not strictly separated.
2) Consider the following picture:

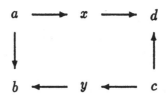

Again, the subtraces $l_1 = (a \to b)$ and $l_2 = (c \to d)$ are not strictly separated and they generate the trace t. Here the letters x and y are between l_1 and l_2.
3) Let t be the trace:

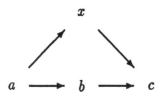

with $l_1 = (a \to b)$, $l_2 = (b \to c)$. Then t is generated by l_1 and l_2, i.e., $t = \langle l_1 \cup l_2 \rangle$, and x is between l_1 and l_2.

Before we continue let us point out a phenomenon which one meets in dealing with trace replacements and which is not present in classical cases of semi-Thue systems and vector replacement systems. Let $(l, r) \in S$ be a rule and $t \underset{(l,r)}{\Longrightarrow} t'$ be a reduction step. If S is a semi-Thue system or a vector replacement system then the value of t' is determined by the replaced subtrace $l \subseteq t$ and by the value of the right-hand side r. (Recall that by the general assumption we do not consider rules (l, r) with $l = 1$.) This is not true anymore if we deal with traces in general. Then the value of t' might depend on the explicit factorization $t = \text{pre}(l)ulv\,\text{suf}(l)$ with $uv = \text{ind}(l)$. To be more precise, let $a \in X$ be a letter which is independent of l, then for $t = al = la$ there is a unique subtrace $l \subseteq t$, but we have $ar \underset{(l,r)}{\Longleftarrow} t \underset{(l,r)}{\Longrightarrow} ra$; and of course $ar \neq ra$, in general. This observation led us in previous papers ([Die87a], [Die89]) to the assumption, called A1), that a and r should commute in all these cases. (Note that the system used in the proof of Theorem 5.3.1 satisfied this assumption.) However, for deciding confluence this assumption is not really necessary. It is enough if the system is confluent on all these pairs (ar, ra). The following lemma is obvious and simply states that the confluence of these pairs is decidable.

Lemma 5.3.4 *Let $S \subseteq M \times M$ be a finite noetherian trace replacement system, $(l, r) \in S$ be a rule, and $a \in X$ be a letter such that a is independent of l. Then it is decidable whether the pair (ar, ra) is confluent.* $\square$

The following considerations are based on the well-founded partial ordering $\prec_S$ of M associated with a noetherian trace replacement system $S \subseteq M \times M$. Recall that for $x, y \in M$ we have $x \prec_S y$ if and only if $y \underset{S}{\overset{*}{\Longrightarrow}} uxv$ for some $u, v \in M$, i.e., x is "smaller" than y if and only if x is a subtrace of some descendant of y. The reason that we do not need the assumption A1) here results from the next lemma.

Lemma 5.3.5 *Let $S \subseteq M \times M$ be a noetherian trace replacement system, $t \in M$ a trace, $l \subseteq t$ a subtrace and $(l, r) \in S$ a rule. Let $u_1 v_1 = u_2 v_2$ be two factorizations of $\text{ind}(l)$ and $t_i = \text{pre}(l)u_i r v_i\,\text{suf}(l)$ for $i = 1, 2$, i.e., $t_1 \underset{(l,r)}{\Longleftarrow} t \underset{(l,r)}{\Longrightarrow} t_2$. Then the following implication holds. If the pair (ar, ra) is confluent for all $a \in X$ such that a is independent of l and if the system S is confluent on all traces $t' \in M$ such that $t' \prec t$ then the pair (t_1, t_2) is confluent, too.*

Proof: For simplification of notation it is convenient to observe first that we may assume $\text{pre}(l) = \text{suf}(l) = 1$. Thus, we have $t = \text{ind}(l)l$ and $t_i = u_i r v_i$ for $i = 1, 2$. If we have $|u_1 u_2| = 0$ then $t_1 = t_2$ and the claim follows. If $|u_1 u_2| > 0$ then we may assume $u_1 = ua$ for some $u \in M, a \in X$ with $(a, l) \in I$. Since (ar, ra) is confluent, the pair $(t_1, urav_1)$ is confluent, too. The pair $(urav_1, t_2)$ is confluent by induction since $|uu_2| < |u_1 u_2|$. The lemma follows since $urav_1 \prec t$.
$\square$

The next theorem says that in view of the Knuth-Bendix completion it is no restriction to assume that all pairs of the form (ra, ar) as mentioned in the lemma above are confluent. This can be achieved by some sort of pre-completion which always terminates on finite systems.

Theorem 5.3.6 *Let $\prec$ be an admissible well-ordering of $M = M(X, D)$ and $S \subseteq M \times M$ be a trace replacement system such that $l \succ r$ for all $(l, r) \in S$. Then there exists an equivalent system $T \subseteq M \times M$ such that $l \succ r$ for all $(l, r) \in T$ and (ra, ar) is confluent by T for all $(l, r) \in T$ and $a \in X$ where a and l are independent. Furthermore, if S is finite then T is also finite and effectively foundable.*

Proof: We obtain the system T by the following procedure. For each $a \in X$ consider all $(l, r) \in S$ such that a and l are independent but (ar, ra) is not confluent. Note that this concernes only rules where $\text{alph}(r)$ is no subset of $\text{alph}(al)$. For all these rules (l, r) add the rules $ra \Rightarrow ar$ $(ar \Rightarrow ra$ respectively) to S. $\square$

Definition: Let $S \subseteq M \times M$ be a noetherian trace replacement system and $(l_1, r_1), (l_2, r_2)$ be rules. Then the set of critical traces $CT(l_1, l_2, S)$ is given by the set of traces $t \in M$ satisfying the following two conditions:

1) The left-hand sides l_1, l_2 are subtraces $l_1 \subseteq t, l_2 \subseteq t$ such that $\langle l_1 \cup l_2 \rangle = t$ and $l_1 \subseteq t, l_2 \subseteq t$ are not strictly separated.

2) For all subtraces $l \subseteq t$ such that $(l, r) \in S$ for some $r \in M$ we have for $i = 1$ or for $i = 2$ that $\langle l_i \cup l \rangle = t$ and $l_i \subseteq t, l \subseteq t$ are not strictly separated.

To illustrate the notion of $CT(l_1, l_2, S)$ consider the semi-Thue case $S \subseteq X^* \times X^*$, where for further simplification S is assumed to be normalized, thus every left-hand side of S is irreducible with respect to all other rules. Then a word $w \in X^*$ belongs to $CT(l_1, l_2, S)$ if and only if $w = l_1 v = u l_2, l_1, l_2$ have over-lapping, and there is no third left-hand-side occuring in w. Thus, the word w gives rise to a minimal critical pair $(r_1 v, u r_2)$.

The following theorem shows that the decidability of confluence can be based on these sets $CT(l_1, l_2, S)$. It is a strengthening of [Die87a, Thm. 2.3] and can be viewed as an analogue to the Winkler-Buchberger criterion for term rewriting systems [WB83], see also [KMN88, section 4,(C1)] and [BD88] for a rather general treatment of critical pair criteria. Compare it also with Theorem 4.1.1

Theorem 5.3.7 *Let $M = M(X, D)$ and $S \subseteq M \times M$ be a noetherian trace replacement system such that (ra, ar) is confluent for all $a \in X$, $(l, r) \in S$ where a and l are independent. Then a set of critical pairs for S is given as follows:*

$$CritPair(S) = \{(t_1, t_2) \mid t_1 \underset{(l_1, r_1)}{\Longleftarrow} t \underset{(l_2, r_2)}{\Longrightarrow} t_2 \text{ for some } t \in CT(l_1, l_2, S),$$

$$(l_i, r_i) \in S, t_i = \mathrm{pre}(l_i) r_i \, \mathrm{ind}(l_i) \, \mathrm{suf}(l_i), i = 1, 2\}$$

Proof: Consider $t_1 \underset{(l_1, r_1)}{\Longleftarrow} t \underset{(l_2, r_2)}{\Longrightarrow} t_2$ for some $t \in M$ such that S is confluent on all $t' \in M$ where $t' \prec_S t$. By Lemma 5.3.5 we may assume that $t_i = \mathrm{pre}(l_i) r_i \, \mathrm{ind}(l_i) \, \mathrm{suf}(l_i)$ for $i = 1, 2$. We have to show that $t \notin CT(l_1, l_2, S)$ implies the confluence of (t_1, t_2). Clearly, we may assume that $t = \langle l_1 \cup l_2 \rangle$ and $l_1 \subseteq t$, $l_2 \subseteq t$ are not strictly separated. Since $t \notin CT(l_1, l_2, S)$, we find a subtrace $l \subseteq t$ for some rule $(l, r) \in S$ such that for $i = 1$ and $i = 2$ we have that $\langle l_i \cup l \rangle \neq t$ or $l_i \subseteq t, l \subseteq t$ are strictly separated. For $i = 1, 2$ we choose any $t'_i \underset{(l_i, r_i)}{\Longleftarrow} t \underset{(l, r)}{\Longrightarrow} t''_i$ such that (t'_i, t''_i) is confluent. This is possible: if $l_i \subseteq t, l \subseteq t$ are strictly separated then we may write $t = ul_ivlw$ ($t = ulvl_iw$ respectively) and the pair (ur_ivlw, ul_ivrw) $((ulvr_iw, urvl_iw)$ respectively) will do, if $\langle l_i \cup l \rangle \neq t$ there exists such a pair since $\langle l_i \cup l \rangle \prec t$. By standard techniques we are reduced to show the confluence of the following pairs $(t_1, t'_1), (t'_1, t''_1), (t''_1, t''_2), (t''_2, t'_2), (t'_2, t_2)$. The confluence of $(t'_1, t''_1), (t'_2, t''_2)$ is known by construction. The confluence of the other three pairs follows by Lemma 5.3.5. $\square$

Directly from the theorem above we obtain the following result:

Corollary 5.3.8 *Let $S \subseteq M \times M$ be a noetherian trace replacement system. Then the system S is confluent if and only if the following two assertions hold:*

i) The pair (ar, ra) is confluent for all $(l, r) \in S, a \in X$ such that $(a, l) \in I$

ii) For all $(l_1, r_1), (l_2, r_2) \in S$ the system S is confluent on the set of critical traces $CT(l_1, l_2, S)$ defined above. $\square$

Remark 5.3.9 Observe that, in general ii) does not imply i), unless M is free or commutative (in which case it does for trivial reasons). Indeed if M is neither free nor commutative then there are three different letters $a, b, c \in X$ such that $(a, c) \in I$ and $(b, c) \in D$. Consider the one-rule system $S = \{a \Longrightarrow b^2\}$. Then $CT(a, a, S) = \{a\}$ and the system is confluent on $\{a\}$ although $(b^2 c, cb^2)$ is not confluent.

We can give a slightly stronger version of the corollary above taking into account that (ar, ra) must be confluent if $al = la$. This version will be used in the last section on so-called cones and blocks. It is asymmetric in the sense that we use the sets $\mathrm{pre}(l_i)$ instead of $\mathrm{suf}(l_i)$. The analogous statement using $\mathrm{suf}(l_i)$ is left to the reader.

Corollary 5.3.10 *Let $S \subseteq M \times M$ be a noetherian trace replacement system. Then the system S is confluent if and only if the following two assertions hold:*

i) The pair (ar, ra) is confluent for all $(l, r) \in S$, $a \in X$ such that $al = la$.

ii) The pair

$$(t_1, t_2) = (\mathrm{pre}(l_1) r_1 \, \mathrm{ind}(l_1) \, \mathrm{suf}(l_1), \mathrm{pre}(l_2) r_2 \, \mathrm{ind}(l_2) \, \mathrm{suf}(l_2))$$

is confluent for all $(l_1, r_1), (l_2, r_2) \in S$ and $t \in CT(l_1, l_2, S)$ such that for $i = 1, 2$ no maximal letter of $\mathrm{pre}(l_i)$ commutes with l_i.

Proof:　Use the same techniques as above. □

The key to our decidability result below is the fact that the sets $CT(l_1, l_2, S)$ are effectively calculable recognizable trace languages. To see this we introduce sets $B(p, q, Y)$. Such a set roughly stands for the set of possible traces between p and q with alphabet Y. Formally $B(p, q, Y)$ is defined for traces $p, q \in M$ and subsets $Y \subseteq X$ by

$$B(p, q, Y) = \{y \in M \mid \quad \mathrm{alph}(y) = Y \text{ and with respect to}$$
$$\text{the trace } pyq \text{ it holds}$$
$$\mathrm{suf}(p) = yq \text{ and } \mathrm{pre}(q) = py\}$$

Note that in the free commutative case $M = \mathbb{N}^X$ the set $B(p, q, Y)$ will be empty unless $Y \subseteq \mathrm{alph}(p) = \mathrm{alph}(q)$. More generally, $B(p, q, Y)$ is empty unless for each maximal element $a \in p$ there exists some $b \in Y \cup \mathrm{alph}(q)$ with $(a, b) \in D$ and for each minimal element $c \in q$ there exists $b \in Y \cup \mathrm{alph}(p)$ with $(b, c) \in D$. If $B(p, q, Y)$ is non-empty then it contains those $y \in M$ with $\mathrm{alph}(y) = Y$ such that every minimal letter of y depends on some letter in p and every maximal letter of y depends on some letter in q. Thus, in all cases $B(p, q, Y)$ is recognizable.

Theorem 5.3.11 *Let $S \subseteq M \times M$ be a noetherian trace replacement system such that the set of left hand sides is recognizable. Then the set $CT(l_1, l_2, S)$ is an effectively calculable recognizable trace language for all $(l_1, r_1), (l_2, r_2) \in S$.*

Proof:　Let $t \in M$ such that $t \in CT(l_1, l_2, S)$. Then $l_1 \subseteq t$, $l_2 \subseteq t$ are non-strictly separated subtraces and we have $t = \langle l_1 \cup l_2 \rangle$. Define the following nine subtraces of t:

$$
\begin{aligned}
p_i &= l_i \cap \mathrm{pre}(l_j) & ,\{i,j\} = \{1,2\}, \\
s_i &= l_i \cap \mathrm{ind}(l_j) = \mathrm{ind}(l_j) & ,\{i,j\} = \{1,2\}, \\
q_i &= l_i \cap \mathrm{suf}(l_j) & ,\{i,j\} = \{1,2\}, \\
s &= l_1 \cap l_2 & \\
y_1 &= \mathrm{suf}(l_1) \cap \mathrm{pre}(l_2) & \\
y_2 &= \mathrm{suf}(l_2) \cap \mathrm{pre}(l_1) &
\end{aligned}
$$

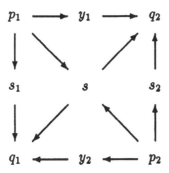

Figure 5.1. A trace $t = \langle l_1 \cup l_2 \rangle$ devided into nine subtraces

We have a picture as in Figure 5.1.
The following formulae hold:

1) $p_i s_i s q_i = l_i$ for $i = 1, 2$

2) $(s_i, l_j) \in I$ for $\{i, j\} = \{1, 2\}$

3) $(p_1, p_2) \in I$, $(q_1, q_2) \in I$

4) $y_i \in B(p_i, q_j, Y_i)$ for some $Y_i \subseteq \{a \in X \mid (a, p_j s_1 s s_2 q_i) \in I\}$, $\{i, j\} = \{1, 2\}$, $Y_1 \times Y_2 \subseteq I$

5) $s \neq 1$ or $p_1, p_2, q_1, q_2 \neq 1$

    Vice versa, if the formulae 1) - 5) hold for some p_i, s_i, s, q_i, y_i and $i = 1, 2$ then $t = p_1 p_2 y_1 s_1 s s_2 y_2 q_1 q_2$ yields a trace with non-strictly separated subtraces $l_1 \subseteq t$, $l_2 \subseteq t$ such that $\langle l_1 \cup l_2 \rangle = t$. Thus, the set $CT(l_1, l_2, S)$ is a subset of a finite union of recognizable sets of the form

$$p_1 p_2 B(p_1, q_2, Y_1) s_1 s s_2 B(p_2, q_1, Y_2) q_1 q_2.$$

    In the following we may think that the data $p_1, p_2, s_1, s, s_2, q_1, q_2 \in M$ and $Y_1, Y_2 \subseteq X$ are fixed. There are only finitely many traces where $p_1 s_1 q_1 = 1$ or $p_2 s_2 q_2 = 1$, thus we assume $p_1 s_1 q_1 \neq 1 \neq p_2 s_2 q_2$. In the next step one replaces $B(p_i, q_j, Y_i)$ by $B_i = \{y \in B(p_i, q_i, Y_i) \mid p_i s_i y, y s_j q_j \in \text{Irr}(S)\}$ for $\{i, j\} = \{1, 2\}$. This is possible without loosing anything from $CT(l_1, l_2, S)$. In fact, say $p_1 s_1 y_1$ is reducible by some rule $(l, r) \in S$. Then the subtrace $l \subseteq p_1 s_1 y_1 \subset t$ is strictly separated from $l_2 \subseteq t$ and $\langle l_1 \cup l \rangle \neq t$ since $p_2 s_2 q_2 \neq 1$. Note that $B = p_1 p_2 B_1 s_1 s s_2 B_2 q_1 q_2$ is recognizable. But it is still too large. It may contain traces

t such that $l \subseteq t$ is a subtrace for some left-hand side where $\langle l_1 \cup l \rangle \neq t \neq \langle l_2 \cup l \rangle$. It is not very difficult to exclude these traces, too, by distinguishing several cases. This is left to the reader, since in our applications the set B is already finite. $\square$

We now state the main result of this section which follows directly from the theorem above together with Lemma 5.3.4.

Corollary 5.3.12 *Let S be a finite noetherian trace replacement system and $L = \{l \in M \mid (l, r) \in S$ for some $r \in M\}$ be the set of left-hand sides. Then it is decidable whether the set*

$$CT(S) = \bigcup_{\{l_1, l_2\} \subseteq L} CT(l_1, l_2, S)$$

is finite. If the set $CT(S)$ is finite then it is decidable whether the system S is confluent. $\square$

Remark 5.3.13 The exact calculation of the set $CT(S)$ above seems to be very difficult in general. However, in order to prove that $CT(S)$ is finite it is enough to prove an upper bound on this set. For example we might prove that the length of traces in $CT(S)$ cannot exceed a certain length. Then we can test the confluence on all traces up to this length without knowing the explicit description of $CT(S)$. $\square$

Example: Consider the same trace as in a previous example:

We may view this as a process consisting of two subprocesses $l_1 = a \to b$ and $l_2 = c \to d$ which may start independently and terminate independently after both have started. Let $S = \{(l_1, r_1), (l_2, r_2)\}$ be a replacement system. Assume that M is generated by $\{a, b, c, d\}$. Then the set of critical pairs as given above has one element only: $\{(ar_2b, cr_1d)\}$. Clearly, it depends on r_1, r_2 whether S is confluent or not; if $r_1 = c$, $r_2 = a$ then S is confluent since $ac = ca$, if $r_1 = da$, $r_2 = bd$ then S is not confluent since $abdb \underset{S}{\Longrightarrow} dadb = dabd \underset{S}{\Longrightarrow} ddad$, $cdad \underset{S}{\Longrightarrow} bdad = dbad$ and these elements are different. This shows also that we have to consider critical pairs, even if the left-hand sides have no overlapping.

Another strange thing happens if there is another generator x in M. Surprisingly this has influence on the set of critical pairs and on the confluence of the system.

Let M be generated by a, b, c, d, and x. There is a trace t' in M such that $l_1 \subseteq t'$, $l_2 \subseteq t'$, l_1 and l_2 are not strictly separated, and some letter is between l_1 and l_2 if and only if $(x, a), (x, d) \in D$ and $(x, b), (x, c) \notin D$. In this case we must have the following picture

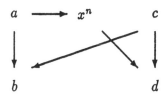

The resulting set of critical pairs is $\{(ax^n r_2 b, cx^n r_1 d) \mid n \geq 0\}$. Again it depends on r_1 and r_2 whether S is confluent or whether actually there is a finite set of critical pairs: If $r_1 = r_2 = 1$ then $ax^n b = abx^n \underset{S}{\Longrightarrow} x^n \underset{S}{\Longleftarrow} x^n cd = cx^n d$. Hence S is confluent and the empty set is a set of critical pairs, too. If $r_1 = a$, $r_2 = c$ then S is not confluent, and it is easy to see that no proper subset of $\{(ax^n cb, cx^n ad) \mid n \geq 1\}$ can be a set of critical pairs for S.

Thus, the system
$$S = \{ab \Rightarrow a, \quad cd \Rightarrow c\}$$
is not confluent on M although it is confluent on the submonoid generated the letters occurring in S.

Later we shall give a sufficient condition that no such situation arises.

Example: Consider two processes $l_1 = a \to b$ and $l_2 = b \to c$. Let $S = \{(l_1, r_1), (l_2, r_2)\}$ be a replacement system having these processes on the left-hand side. Again, the same phenomena as above arises. If M is generated by $\{a, b, c\}$ then a set of critical pairs is given by $\{(r_1 c, ar_2)\}$. Assume there is another generator x. Let $t \in M$ be a trace with $l_1 \subseteq t$, $l_2 \subseteq t$, $t = \langle l_1 \cup l_2 \rangle$, and l_1, l_2 are not strictly separated. Say there is a letter between l_1 and l_2. Then it holds $(x, a), (x, c) \in D$, but $(x, b) \notin D$ and every such trace t has the form

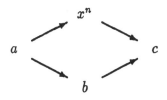

It follows that the set $\{(r_1 x^n c, a x^n r_2) \mid n \geq 0\}$ is a set of critical pairs for S.
□

So far, in the computation of critical pairs mainly left-hand sides have been taken into acount. A simplification arises if we put some restriction on the possible right-hand sides. Consider the case where we have $ra = ar$ whenever the letter a is independent of any letter in the alphabet of the left-hand side l for $(l, r) \in S$. This is satisfied for example in the semi-Thue case, in the case of vector replacement systems or if r is central in M. Hence, in particular in the case of special systems, i.e., where the right-hand sides are all equal to 1. Then it is enough to consider $t_1 \underset{(l_1, r_1)}{\Longleftarrow} t \underset{(l_2, r_2)}{\Longrightarrow} t_2$ where $t = \langle l_1 \cup l_2 \rangle$ and l_1, l_2 have overlapping, i.e., $l_1 \cap l_2 \neq \emptyset$. Furthermore the resulting critical pair (t_1, t_2) has a simpler description where the parts between l_1 and l_2 become pre- and suffixes.

Theorem 5.3.14 *Let $S \subseteq M \times M$ be a noetherian trace replacement system over $M = M(X, D)$. Assume that every letter which is independent of any letter of a left-hand side of a rule commutes with the corresponding right-hand side. Then the system S is confluent if and only if the system is confluent on the set of traces $t \in M$ which satisfy the following two conditions:*

i) For some $(l_1, r_1), (l_2, r_2) \in S$ there are subtraces $l_1 \subseteq t$, $l_2 \subseteq t$ such that $t = \langle l_1 \cup l_2 \rangle$, $l_1 \cap l_2 \neq \emptyset$.

ii) If $l \subseteq t$ is any subtrace for some rule (l, r) then for $i = 1$ or $i = 2$ we have $t = \langle l_i \cup l \rangle$ and $l_i \cap l \neq \emptyset$.

Furthermore, for each such trace t let $p_i = \mathrm{pre}(l_j) \cap l_i$, $s_i = \mathrm{ind}(l_j)$, $q_i = \mathrm{suf}(l_j) \cap l_i$, $y_i = (\mathrm{suf}(l_i) \cap \mathrm{pre}(l_j))$, $\{i, j\} = \{1, 2\}$, and $s = l_1 \cap l_2$. Then for the resulting critical pair $t_1 \underset{(l_1, r_1)}{\Longleftarrow} t \underset{(l_2, r_2)}{\Longrightarrow} t_2$ we have the formula:

$$(t_1, t_2) = (y_1 p_2 r_1 s_2 q_2 y_2, y_2 p_1 r_2 s_1 q_1 y_1)$$

Proof: First, observe that the assumption on S implies that $ra = ar$ for all $a \in X$, $(l, r) \in S$ where a and l are independent. Thus, we may apply Theorem 5.3.7. Now, consider $t_1 \underset{(l_1, r_1)}{\Longleftarrow} t \underset{(l_2, r_2)}{\Longrightarrow} t_2$ with $t = \langle l_1 \cup l_2 \rangle$. Define the nine subtraces p_1, p_2, s, s_1, s_2, q_1, q_2, y_1, y_2 as above. Then we may write $t_i = p_j y_j s_j r_i y_i q_j$ for $\{i, j\} = \{1, 2\}$. Since we may assume that l_1, l_2 are not strictly separated the assumption on S implies the commutation of r_i with y_i and y_j. Using the other independencies we obtain $t_i = y_i p_j s_j r_i q_j y_j$. This is already the formula above. Following the proof of Theorem 5.3.7 it is enough to show that (t_1, t_2) is confluent if $l_1 \cap l_2 = \emptyset$, i.e., if $s = 1$. So, let $s = 1$ then we have $p_1 \neq 1$, $p_2 \neq 1$, $q_1 \neq 1$, $q_2 \neq 1$ because otherwise l_1, l_2 would be strictly separated in t. Then the assumption on S says $p_j s_j r_i = r_i p_j s_j$ and $r_i q_j = q_j r_i$ for $\{i, j\} = \{1, 2\}$. Hence $(t_1, t_2) = (y_1 r_1 p_2 s_2 q_2 y_2, y_2 p_1 s_1 q_1 r_2 y_1)$. Since $s = 1$ we have $(t_1, t_2) = (y_1 r_1 l_2 y_2, y_2 l_1 r_2 y_1)$. This pair reduces to $(y_1 r_1 r_2 y_2, y_2 r_1 r_2 y_1)$. However, we have $y_1 r_1 r_2 y_2 = r_1 r_2 y_1 y_2 = y_2 r_1 r_2 y_1$ and this concludes the proof. $\square$

It is open whether the structure of critical pairs given in the theorem above is simple enough to decide the confluence. Some further simplifications occur if the system is special, i.e., if all right-hand sides are equal to 1. But even for finite special trace replacement systems we were not able to solve the question of confluence in general. Of course if the subtraces y_1, y_2 are empty then confluence is decidable. A nice example for this is given by the presentation of free partially commutative groups.

Example: Let (X, D) be a dependence alphabet and $M = M(X, D)$. Let $\overline{X}$ be a disjoint copy of X and set $\tilde{X} = X \dot\cup \overline{X}$. Define on $\tilde{X}$ a dependence relation $\tilde{D} \subseteq \tilde{X} \times \tilde{X}$ by $(\tilde{x}, \tilde{y}) \in \tilde{D}$ if $(i(\tilde{x}), i(\tilde{y})) \in D$ where $i : \tilde{X} \to X$ denotes the canonical mapping $i(x) = i(\overline{x}) = x$ for $x \in X$.

Let $\tilde{S} \subseteq M(\tilde{X}, \tilde{D}) \times M(\tilde{X}, \tilde{D})$ be the special replacement system which is given by the following rules: (1): $x\overline{x} \Longrightarrow 1$, (2): $\overline{x}x \Longrightarrow 1$, for $x \in X$.

Then $\tilde{X}^* / \tilde{S}$ is the free partially commutative group over (X, D). This is also the group associated with $M(X, D)$. It is very easy to see that the restrictions put on the critical pairs according to the last corollary imply that critical pairs arise only as follows

$$x \underset{(1)}{\Longleftarrow} x\overline{x}x \underset{(2)}{\Longrightarrow} x, \qquad \overline{x} \underset{(2)}{\Longleftarrow} \overline{x}x\overline{x} \underset{(1)}{\Longrightarrow} \overline{x}.$$

Hence, $\tilde{S}$ is a complete system. It may be used for normal forms of elements of free partially commutative groups. Since $M \subseteq \mathrm{Irr}(\tilde{S}) \subseteq M(\tilde{X}, \tilde{D})$, we obtain the well-known fact that free partially commutative monoids are embeddable in groups.

5.4 The Completion Procedure on Traces

In the second section we described the Knuth-Bendix completion procedure in general. We could apply it to our situation of free partially commutative monoids if there would be an effective way to compute finite sets of critical pairs. But we have already seen that this is an impossible task. We could restrict ourselves to systems $S \subseteq M \times M$ where the recognizable set $CT(S) \subseteq M$ of critical traces is finite and stays finite during the completion procedure. However, the computation of $CT(S)$ and hence of $CritPair(S)$ can be very difficult. So, a calculation of this set should be avoided whenever possible. We will see that indeed the calculation of $CT(S)$ can be postponed at the end of the procedure.

The idea is to use during the procedure the following sets which we have called *basic critical pairs*.

Definition: Let $S \subseteq M \times M$ be a trace replacement system over $M = M(X, D)$ and let $I \subseteq M \times M$ be the independence relation. The set of $BCP(S)$ of basic critical pairs for S is given by the following formula

$$
\begin{aligned}
BCP(S) \;=\; & \{(ra, ar) \mid a \in X, (l, r) \in S, (a, l) \in I, ra \neq ar\} \\
& \cup \{(t_1, t_2) \mid \exists (l_1, r_1), (l_2, r_2) \in S, t \in M \text{ with} \\
& \quad \text{subtraces } l_1 \subseteq t, l_2 \subseteq t \text{ such that} \\
& \quad l_1, l_2 \text{ are not strictly separated, every} \\
& \quad \text{letter of } t \text{ belongs to } l_1 \text{ or to } l_2, \text{ and} \\
& \quad t_1 \underset{(l_1, r_1)}{\Longleftarrow} t \underset{(l_2, r_2)}{\Longrightarrow} t_2\} \qquad \square
\end{aligned}
$$

It is clear that the set $BCP(S)$ is finite if $S \subseteq M \times M$ is finite and in the next section we will see that $BCP(S)$ can be computed in polynomial time in the size of the system S. A more explicit description of $BCP(S)$ can be easily derived from the proof of Theorem 5.3.11 observing that we consider only the cases where $y_1 = \mathrm{pre}(l_2) \cap \mathrm{suf}(l_1) = y_2 = \mathrm{pre}(l_1) \cap \mathrm{suf}(l_2) = \emptyset$. The following remark says that basic critical pairs are in some sense the bottom of all cirtical pairs.

Remark 5.4.1 Let $\prec$ be an admissible well-ordering of a trace monoid M and $S \subseteq M \times M$ be a trace replacement system. Then for all $t_1 \underset{S}{\Longleftarrow} t \underset{S}{\Longrightarrow} t_2$ with $t_1 \neq t_2$ there exists a basic critical pair $(t'_1, t'_2) \in BCP(S)$ and some $t' \preceq t$ such that $t'_1 \underset{S}{\Longleftarrow} t' \underset{S}{\Longrightarrow} t'_2$

Proof: Observe first that for all $t \in M$ and all subsets $l \subseteq t$, $l \neq t$ (not necessarily subtraces) we have $l \prec t$. To see this we reduce by induction to the case $t \setminus l = \{a\}$. Then $t = l'al''$ for some $l'l'' = l$ and since $1 \prec a$ we have $l \prec t$. The proof follows easily from this observation. $\square$

We will run a modified completion procedure as follows. We choose an admissible well-ordering $\prec$ and we start with a finite system S_1 where $l \succ r$ for all $(l, r) \in S_1$. We proceed as usual but for the sets $C(S_i)$ we always take the finite sets $BCP(S_i)$ of basic critical pairs. According to the completion procedure we set $S^* = \bigcup_{i \geq 1} S_i$. Of course, there is no algorithmic way to decide whether the completion procedure stops or not. We are mainly interested in the case where S^* is finite, i.e., at a certain point in the procedure we arrive the situation that all basic critical pairs are confluent. The unsolvable problem is that in general we can not say whether the finite system S^* is confluent or not. (Otherwise we would obtain a contradiction to the undecidability result of Narendran/Otto,

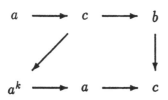

Figure 5.2. The trace $ac(a^{k+1} + b)c$ for $k \geq 0$

Theorem 5.3.2).We only can try to prove the confluence by some sufficient condition. For example, we show that $CT(S^*)$ is finite. If $CT(S^*)$ is finite then we can test the confluence of S^* on this set. Note that the test will be positive and can therefore be omitted if we have $CritPair(S^*) = BCP(S^*)$. Because the calculation of $CT(S^*)$ is difficult, we should however normalize S^* first. Since S^* is not known to be complete we have to be a little bit careful with the normalization: We reduce the right-hand sides rule after rule until they are irreducible. Call this new system S^* again. Now, if $(l,r) \in S^*$ and $l \notin Irr(S^* \setminus \{(l,r)\})$ then compute an irreducible descendant $\hat{l}$ of l with respect to $S^* \setminus \{(l,r)\}$. If $\hat{l} = r$ then we safely may omit the rule (l,r), otherwise S^* is not complete and we restart the whole procedure again with the system $S^* \setminus \{(l,r)\}$ and the new rule $(\hat{l},r)$ (or $(r,\hat{l})$ if $r \prec \hat{l}$). If, finally, S^* is normalized, confluent on $BCP(S^*)$ and for some further reasons we know that $BCP(S^*) = CritPair(S^*)$, then S^* is a finite normalized complete system for M/S_1.

In the following examples we show that this might be a feasible way to treat the completion procedure on traces. We use the notational convention to allow plus-signs between letters which commute. We always use well-orderings as in Proposition 5.1.6 or 5.1.8 without mentioning them explicitly.

Example: Let $M = \{a, b, c\}^* / \{(ab, ba)\}$, the dependence graph of M is $a - c - b$. Consider the following replacement system

$$S_1 = \{((a + b)c, 1), (bca, 1), (acb, 1)\}.$$

For this system the set of critical traces $CT(S_1)$ is infinite. Indeed, it contains all traces of the form $t = ac(a^{k+1} + b)c$, $k \geq 0$ as shown in Figure 5.2.

However, as we have said above we concentrate on basic critical pairs. Thus we obtain $BCP(S_1) = \{(ca, ac), (cb, bc)\}$. Assume our well-ordering says $ac \prec ca$ and $bc \prec cb$. Then, after normalization we have

$$S^* = S_2 = \{((a + b)c, 1), (ca, ac), (cb, bc)\}.$$

Now, let us calculate $CT(S^*)$. We see that all traces $ac(a^{k+1} + b)c$ with $k+1$ vanish from $CT(S^*)$ since ca is now a left-hand side of rule. It easily follows that

$CT(S^*) = \{(a+b)ca, (a+b)cb, c(a+b)\}$. The set $CT(S^*)$ yields the basic critical pairs and these are confluent. The unique normal forms are $(a^2+b)c$, $(a+b^2)c$, $(a+b)c$ respectively. Thus, S^* is a finite complete replacement system for M/S_1. We have $M/S_1 = M/S^* \cong \mathbf{Z} \times \mathbf{Z}$.

It is worth noting that $\mathbf{Z} \times \mathbf{Z}$ is not presentable by any finite complete semi-Thue system over a free monoid on three letters, see [Die86b]. Here, one commutation rule, namely $ab = ba$, is enough to obtain such a complete system on three generators.

We finish this section with an non-commutative example which is therefore not presentable by any vector-replacement system, and where the Knuth-Bendix completion algorithm would not terminate in the purely free case neither.

Example: Let $M = \{a, b, c, g\}^* / \{(ab, ba), (ac, ca), (bc, cb)\}$
The monoid M is represented by its dependence graph:

It is isomorphic to $\mathbf{N}^3 * \mathbf{N}$. We use a well ordering on M as it is given in Proposition 5.1.8.

Consider the following finite noetherian system
$S_1 = \{((a+b+c), 1), (g^2, 1), (ga, bg)\}$.
The system S_1 is not complete. We have
$bg(b+c) \underset{S_1}{\Longleftarrow} g(a+b+c) \underset{S_1}{\Longrightarrow} g, \; gbg \underset{S_1}{\Longleftarrow} g^2 a \underset{S_1}{\Longrightarrow} a$.
According to the completion procedure we obtain:
$S_2 = \{(a+b+c, 1), (g^2, 1), (ga, bg), (bg(b+c), g), (gbg, a)\}$. This system is not complete since
$gb \underset{S_2}{\Longleftarrow} gbg^2 \underset{S_2}{\Longrightarrow} ag$. After the next step, and normalization, we obtain:
$S_3 = \{(a+b+c, 1), (g^2, 1), (ga, bg), (gb, ag), ((a+b)gc, g)\}$.
Again, S_3 is not complete. Consider:
$gc \underset{S_3}{\Longleftarrow} (a+b+c)gc \underset{S_3}{\Longrightarrow} cg$.
But in the next step the completion procedure stops and normalization yields the following system

$$S^* = \{(a+b+c, 1), (g^2, 1), (ga, bg), (gb, ag), (gc, cg)\}.$$

It is straightforward to see that $CT(S^*)$ is finite and that $CritPair(S^*) = BCP(S)$. The confluence of S^* is can be proved by inspecting all critical pairs.

It follows that S^* is a complete replacement system. (We may use this system to solve the word problem for M/S_1 in linear time!) The algebraic structure of M/S_1 is given by the semi-direct product of $\mathbf{Z} \times \mathbf{Z}$ by $\mathbf{Z}/2\mathbf{Z}$ where $\mathbf{Z}/2\mathbf{Z}$ operates on $\mathbf{Z} \times \mathbf{Z}$ via the matrix $\begin{pmatrix} 0 & 1 \\ 1 & 0 \end{pmatrix} \in SL_2(\mathbf{Z})$.

5.5 Some Remarks on Complexity

In this section we prove some complexity results. First we give an upper bound for the number of basic critical pairs which are computed according to the definition of the preceeding section. The notations are as usual:

Let $M = M(X, D)$ be a trace monoid generated by some finite alphabet X and $k \geq 0$ the least integer such that M has an embedding into a k-fold direct product of free monoids, $M \longrightarrow \prod\limits_{i=1}^{k} X_i^*$.

Let $S \subseteq M \times M$ be a finite replacement system of n rules, $n \geq 1$ and let $m := \max\{|l| \mid (l,r) \in S\}$. The size $||S||$ of S is defined by $||S|| := \sum\limits_{(l,r) \in S} (|l| + |r|)$.

Theorem 5.5.1 *The set of basic critical pairs*

$$
\begin{aligned}
BCP(S) \;=\; & \{(ra, ar) \mid a \in X, (l,r) \in S, (a,l) \in I, ra \neq ar\} \\
& \cup \{(t_1, t_2) \mid \exists (l_1, r_1), (l_2, r_2) \in S, t \in M \text{ with} \\
& \qquad \text{subtraces } l_1 \subseteq t, l_2 \subseteq t \text{ such that} \\
& \qquad l_1, l_2 \text{ are not strictly separated, every} \\
& \qquad \text{letter of } t \text{ belongs to } l_1 \text{ or to } l_2, \text{ and} \\
& \qquad t_1 \underset{(l_1, r_1)}{\Longleftarrow} t \underset{(l_2, r_2)}{\Longrightarrow} t_2 \} \qquad \square
\end{aligned}
$$

contains at most $\dfrac{4^k (m+1)^k n(n+1)}{2} + n(\#X - 1)$ *elements.*

Proof: Clearly, there are at most $n(\#X - 1)$ pairs of the form (ra, ra) such that $ra \neq ar$, $(l,r) \in S$, $a \in X$, and $(a,l) \in I$. Hence it is enough to consider the other pairs of $BCP(S)$. Using the embedding $M \longrightarrow \prod\limits_{i=1}^{k} X_i^*$ every trace $t \in M$ becomes a k-tupel of words $t = (t_1, \ldots, t_k)$. Let $(p,r), (q,s) \in S$ be rules, we count the number of traces t with $t = p \cup q$ with multiplicities according to the different ways how p and q may occur as subtraces $p \subseteq t$, $q \subseteq t$.

Since $t = p \cup q$, every letter of t is in $p \subseteq t$ or in $q \subseteq t$. Hence, for all $i = 1, \ldots, k$ there are $u_i, v_i, w_i \in X_i^*$: $t_i = u_i v_i w_i$ with

	1)	$u_i v_i = p_i$	and	$v_i w_i = q_i$
or	2)	$u_i v_i = q_i$	and	$v_i w_i = p_i$
or	3)	$u_i v_i w_i = p_i$	and	$v_i = q_i$
or	4)	$u_i v_i w_i = q_i$	and	$v_i = p_i$

Thus, for each $i \in \{1, \ldots, k\}$ there are at most $4(m+1)$ choices for $u_i, v_i, w_i \in X_i^*$. For each pair (p, r), $(q, s) \in S$ this gives $4^k (m+1)^k$ combinations. The result follows because we have to consider at most $\frac{n(n+1)}{2}$ pairs of rules. $\square$

Note that the theorem above is a roughly estimated worst-case analysis. In practice, there should be much less basic critical pairs.

Corollary 5.5.2 *For fixed M there is a polynomial time algorithm (in the input size $\|S\|$) which computes the set of basic critical pairs $BCP(S)$.*

Proof: Let (p, r), $(q, s) \in S$ be rules. In polynomial time we compute all k-tuples of words $(t_1, \ldots, t_k) = (u_1, \ldots, u_k)(v_1, \ldots, v_k)(w_1, \ldots, w_k)$ which satisfy one of the conditions 1), ..., 4) in the proof of Theorem 5.5.1. From these tuples we obtain factorizations $t = xpy = x'qy'$ in $X_1^* \times \ldots \times X_k^*$. Then one verifies that x, x', y represent traces in linear time, see Sect. 1.5. (If x, x', y are traces, then t is a trace and p, q are subtraces.) Next, one decides whether $p \cap q \neq \emptyset$ or whether p, q are not strictly separated, which is simple for tuples. Finally, one computes the critical pairs from the above factorizations. $\square$

A replacement system $S \subseteq M \times M$ over a trace monoid M is called *recognizable* if the number of right-hand sides is finite and if for each $r \in M$ the set $\{l \in M \mid (l, r) \in S\}$ is recognizable in M. Recognizable systems are of interest here since there are fast algorithms to compute direct descendants:

Lemma 5.5.3 *Let $S \subseteq M \times M$ be a recognizable trace replacement system. Then there is an algorithm which on input $t \in M$ decides in time $O(|t|)$ whether t is reducible with respect to S. If it is then it computes a direct descendant of t.*

Proof: This follows from Corollary 2.2.3. $\square$

For finite systems, however, the use of Corollary 2.2.3 in the proof above seems to be not adequate. For finite systems a better algorithm would be based on the function $\text{pre}(l, t)$ described in Sect. 1.5.

Let us measure the time complexity to compute irreducible forms for finite noetherian systems $S \subseteq M \times M$ in terms of the following function

$$d_S : \mathbb{N} \to \mathbb{N}, \quad d_S(n) = \max\{m \in \mathbb{N} \mid t \xRightarrow[S]{m} \hat{t}, \ |t| = n\}$$

This means $d_S(n)$ is the maximal number of possible reduction steps starting on a trace of length n. For applications, one is mainly interested in cases where d_S grows slowly. This is for example the case when $S \subseteq M \times M$ is weight-reducing, then d_S is a linear function.

Theorem 5.5.4 *Let $S \subseteq M \times M$ be a finite noetherian trace replacement system. Then we may compute irreducible normal forms on some (multitape) Turing machine in time $O(d_S^2)$.*

Proof: Consider a derivation chain $t \underset{S}{\Longrightarrow} t_1 \underset{S}{\Longrightarrow} \ldots \underset{S}{\Longrightarrow} t_m \in \mathrm{Irr}(S)$. For some constant $c \geq 0$ we have $|t_i| = |t| + c \cdot i$ for $1 \leq i \leq m$. From the lemma above we can compute the irreducible descendant t_m in time $O(m \cdot |t| + c\frac{m(m-1)}{2})$. Of course, $m \leq d_S(|t|)$ and since we may assume $S \neq \emptyset$ we have $|t| \in O(d_S(|t|))$. Hence, we obtain the estimate above. $\square$

If in the proof above the system S has no length-increasing rule then we can choose the constant c to be zero. We obtain a slightly better timebound. Also we can use any weight function instead of the length. Therefore we can formulate the following.

Corollary 5.5.5 *Let $S \subseteq M \times M$ be a finite complete trace replacement system over $M = M(X, D)$. Then the word problem $s \underset{S}{\overset{*}{\Longleftrightarrow}} t$ for traces $s, t \in M$ can be decided in time $O(d_S^2(|s|) + d_S^2(|t|))$. If for some weight function $\gamma : X \to \mathbb{N}^+$ the system is not weight-increasing then it can be decided in time $O(|s|d_S(|s|) + |t|d_S(|t|))$, and if S is weight-reducing then it can be decided in square time $O(|st|^2)$. $\square$*

Remark 5.5.6 We have no idea whether the time bounds stated above are optimal. On the other hand in the restricted case of semi-Thue systems (or vector-replacement systems) we could obtain the same assertion as in Corollary 5.5.5 in time $O(d_S)$, see Sect. 4.1. Unfortunately the technique of [Boo82] does not apply to the free partially commutative case in general. It applies to the special systems $\tilde{S} \subseteq \tilde{X} \times \tilde{X}$ of Sect. 5.3 which describe free partially commutative groups. It has been shown by C. Wrathall, [Wra88] that these systems allow a linear time algorithm. In the next section we present a more general condition which allows to compute irreducible normal forms in time $O(d_S)$.

5.6 The Condition $G_k(S)$ for $k \geq 0$

Consider a finite noetherian trace replacement system $S \subseteq M \times M$. The informal reason why, so far, we can not prove a better time bound than $O(d_S^2)$ is that for some $c > 0$ there might be irreducible traces $t \in \text{Irr}(S)$ such that if we multiply t by a letter $a \in X$ from the left (or right) then at (or ta) becomes reducible by some rule $(l, r) \in S$, but for all factorizations $at = ulv$, we have $|u| \geq c|t|$ and $|v| \geq c|t|$. Even if there would be, at this stage, a fast way (constant time) to compute the reduction step $at = ulv \underset{S}{\Longrightarrow} urv$, we see no fast way to test whether urv is irreducible, it will take time linear to the length of $|t|$. Of course, all these problems vanish if we could bound the length of $|u|$ in the situation above by some constant $k \geq 0$ depending on S only. This is exactly what the following condition says.

Definition: Let $k \geq 0$, $X^{(k)} = \{t \in M \mid |t| \leq k\}$, and $S \subseteq M \times M$ be a *noetherian trace replacement system with the set of left-hand sides* $L = \{l \in M \mid (l, r) \in S$ for some $r \in M\}$. *We say that the condition* $G_k(S)$ *holds if we have*

$$X \, \text{Irr}(S) \subseteq \text{Irr}(S) \cup X^{(k)} LM.$$

The properties we are interested in are summarized in the next theorem.

Theorem 5.6.1 *Let* $k \geq 0$ *and* $S \subseteq M \times M$ *be a finite noetherian trace replacement system and* L *be the set of left-hand sides. Then we have the following assertions.*

i) *It is decidable whether* $G_k(S)$ *holds.*

ii) *If* $G_k(S)$ *holds then we can decide whether* S *is confluent.*

iii) *If* $G_k(S)$ *holds and if* S *is confluent then we can decide the word problem of* S *in time* $O(d_S)$

Proof: i) trivial since all sets involved for deciding $G_k(S)$ are recognizable.
ii) We show that the set $CT(l_1, l_2, S)$ introduced in Sect. 5.3 is finite for all $l_1, l_2 \in L$ where as above $L = \{l \in M \mid (l, r) \in S$ for some $r \in M\}$. The result then follows by Corollary 5.3.8.

Let $l_1, l_2 \in L$ and $t \in M$ be a trace with subtraces $l_1 \subseteq t$, $l_2 \subseteq t$ such that $\langle l_1 \cup l_2 \rangle = t$ and $l_1 \subseteq t$, $l_2 \subseteq t$ are not strictly separated. As in the proof of Theorem 5.3.11 we devide t into nine subtraces:

$$
\begin{aligned}
p_i &= l_i \cap \text{pre}(l_j) & , && \{i, j\} &= \{1, 2\}, \\
s_i &= l_i \cap \text{ind}(l_j) & , && \{i, j\} &= \{1, 2\}, \\
q_i &= l_i \cap \text{suf}(l_j) & , && \{i, j\} &= \{1, 2\}, \\
s &= l_1 \cap l_2, & && & \\
y_i &= \text{suf}(l_i) \cap \text{pre}(l_j) & , && \{i, j\} &= \{1, 2\}.
\end{aligned}
$$

Then we have $t = p_1 p_2 y_1 s_1 s s_2 y_2 q_1 q_2$.

Let F be the finite set of subtraces of traces in $X^{(k)}L$, i.e., $F = \{f \in M \mid \exists u, v \in M : ufv \in X^{(k)}L\}$. (In fact, for fixed k the cardinality of F is even a polynomial in the size of the system S.) It will be enough to show that if $t \in CT(l_1, l_2, S)$ then we have $y_1 s_2 q_2, y_2 s_1 q_1 \in F$. We do this by contradiction assuming $t \in CT(l_1, l_2, S)$ and $y_1 s_2 q_2 \notin F$. (The case $y_2 s_1 q_1 \notin F$ is symmetric.) Since $s_2 q_2$ is a subtrace of l_2 we have $s_2 q_2 \in F$ and hence $|y_1| \geq 1$. Therefore $p_1 \neq 1$. This implies $y_1 s_2 q_2 \in \text{Irr}(S)$, because otherwise t would contain a subtrace $l \subseteq y_1 s_2 q_2 \subseteq t$ for some $(l, r) \in S$ such that $l_1 \subseteq t$, $l \subseteq t$ are strictly separated and $\langle l_2 \cup l \rangle \neq t$. On the other hand $p_2 s y_1 s_2 q_2 = y_1 l_2$ is reducible since $l_2 \in L$. Hence for some factorization $p_2 s = w'aw$, $w', w \in M$, $a \in X$, we have $wy_1 s_2 q_2 \in \text{Irr}(S)$ and $awy_1 s_2 q_2 \notin \text{Irr}(S)$. It follows from $G_k(S)$ that $awy_1 s_2 q_2 = ulv$ for some $(l, r) \in S$, $u, v \in M$ with $ul \in X^{(k)}L$. Now $y_1 s_2 q_2 \notin F$ implies that $y_1 s_2 q_2$ is no subtrace of ul. Hence one (maximal) letter of $y_1 s_2 q_2$ belongs to $v \subseteq t$, hence $\langle l_1 \cup l \rangle \neq t$. Since $p_1 \neq 1$ we also have $\langle l_2 \cup l \rangle \neq t$, thus we have $t \notin CT(l_1, l_2, S)$.

iii) The following algorithm right_reduce$_k$ is a straightforward generalization of the technique to compute irreducible normal forms for semi-Thue systems, see Sect. 4.1. We simply take the value of k into account in looking for a possible reduction step.

```
function right_reduce_k (s:trace):trace
var t:trace:=1
while s ≠ 1
do {"loop-invariant: t is irreducible, st is congruent to the input"}
    choose some maximal letter a of s;
    s := sa⁻¹; t := at;
    for all (l,r) ∈ S
    do if t ∈ ulM for some u ∈ M, |u| ≤ k
        then s := sur; t := (ul)⁻¹t
        endif
    endfor
endwhile
return t
endfunction
```

It is easy to see that for every noetherian trace replacement system the algorithm terminates on input length n in time $O(d_s(n))$. If in addition the system satisfies $G_k(S)$ then the loop invariant holds and therefore right_reduce$_k$ computes irreducible descendants. Assertion iii) follows. $\square$

In the proof above we used for each $k \geq 0$ a different algorithm. In the next section we present a uniform algorithm which works for every noetherian system S and which has the same time bound $O(d_S)$ if $G_k(S)$ is satisfied for some $k \geq 0$. This is clearly a big advantage. The problem of the new algorithm is however that we need a control structure by finite asynchronous automata. This means we have to use a very complex mechanism which leads to high constants.

Open problem Is it decidable whether for given finite $S \subseteq M \times M$ there exists $k \geq 0$ such that $G_k(S)$ holds? This question would have an affirmative answer if we could decide for given recognizable trace languages $A, B \subseteq M$ with $A \subseteq MB$ whether there exists a finite set $F \subseteq M$ such that $A \subseteq FB$. This seems to be an interesting question for recognizable trace languages independent of trace rewriting. It can be solved for regular word languages.

The reason to consider the condition $G_k(S)$ for different $k \geq 0$ follows from the next proposition:

Proposition 5.6.2 Let $k \geq 0$ and M be neither free nor commutative then there exists a length-reducing trace replacement system $S \subseteq M \times M$ of at most two rules such that $G_{k+1}(S)$ holds but not $G_k(S)$.

Proof: Since M is neither free nor commutative we find three different letters $a, b, c \in X$ such that $(a, b) \in D$ and $(a, c) \in I$. Consider the system $S = \{cb \Longrightarrow 1, a^{k+2} \Longrightarrow 1\}$. Then $t = ca^{k+1}b \in X \operatorname{Irr}(S)$ but neither $t = a^{k+1}cb \in \operatorname{Irr}(S)$ nor $a^{k+1}cb \in X^{(k)}\{cb, a^{k+2}\}M$. Hence $G_k(S)$ does not hold whereas it is easy to see that $G_{k+1}(S)$ is true. $\square$

Remark 5.6.3 i) If M is free or commutative then $G_0(S)$ holds for every system $S \subseteq M \times M$. We will see in the last section that for one-rule systems $G_k(S)$ implies $G_0(S)$ for any $k \geq 0$. Therefore, the proposition above is tight.
ii) In [Ott89] another decidable and sufficient condition is given such that the confluence of finite noetherian trace replacement systems becomes decidable. The approach of Otto is based on the notion of convergence and coherence for term rewriting systems which was developped by Jouannaud, [Jou83]. Inspecting Otto's condition it turns out to be equivalent with confluence in our sense and $G_0(S)$. Since $G_0(S)$ implies $G_k(S)$ for all $k \geq 0$ and any $G_k(S)$ implies the finiteness condition given in Corollary 5.3.12, but non of these implications is reversible, our condition is clearly weaker. Furthermore our approach has the advantage of staying entirely in the theory of traces.

5.7 An Efficient Algorithm for Computing Irreducible Normal Forms

In this section we present an algorithm which computes always irreducible normal forms in time $O(d_S^2)$ but which has the property that whenever the system satisfies $G_k(S)$ for some $k \geq 0$ then it works in time $O(d_S)$. From this viewpoint it is the best known algorithm in this area. The implementation of the algorithm depends essentially on the existence of finite asynchronous automata which were introduced in [Zie87]. The proof that every recognizable trace language is accepted by such automaton has been quite involved, see Sect. 2.4 and the resulting constants are extremely high. So, in practice it might be necessary to work with "less optimal" algorithms. But may be a better understanding of asynchronous automata will change the situation.

The algorithm we are going to construct is based on a notion of protocols which is available for asynchronous automata, but not for usual finite M-automata. In fact, we shall use asynchronous cellular automata. It is done here to have smaller state sets. Let us briefly recall the definition.

A finite M-automaton $U = (Z, \delta, q_0, F)$ (where Z denotes the finite state set, $\delta : Z \times M \to Z$ is the (partially defined) transition mapping, $q_0 \in Z$ is the initial state, and $F \subseteq Z$ is the set of final states) has been called *asynchronous cellular* if the following conditions hold:

1) The state set Z is a cartesian product $Z = \prod_{x \in X} Z_x$

2) The partially defined transition mapping δ is given by a collection of partial mappings
$$\{\delta_a : (\prod_{b \in D(a)} Z_b) \to Z_a \mid a \in X\}$$
where $D(a) = \{b \in X \mid (a, b) \in D\}$ for $a \in X$.

3) For all traces $t \in M$ the state $\delta(q_0, t)$ is defined.

Condition 2) means that for all $a \in X$ and $(z_x)_{x \in X} \in \prod_{x \in X} Z_x$ the next-state $\delta((z_x)_{x \in X}, a)$ is defined if and only if $\delta_a((z_b)_{b \in D(a)})$ is defined. In this case we have $\delta((z_x)_{x \in X}, a)_y = z_y$ for $y \neq a$ and $\delta((z_x)_{x \in X}, a)_a = \delta_a((z_b)_{b \in D(a)})$. Condition 3) is included for technical reasons only.

Let $U = (\prod_{a \in X} Z_a, \delta, q_0, F)$ be a asynchronous cellular automaton. A *protocol* of U is an element of the product space $\prod_{a \in X}(Z_a^+)$ which are inductively defined as follows:

The element $q_0 \in \prod_{x \in X} Z_x \subseteq \prod_{x \in X}(Z_x^+)$ is a protocol. If $p \in \prod_{x \in X}(Z_x^+)$ is a protocol and $a \in X$ is a letter then the protocol ap is defined as follows: Take $q \in \prod_{x \in X} Z_x$ such that $p = qp'$ for some $p' \in \prod_{x \in X} Z_x^*$. Then compute

$\delta(q, a)_a \in Z_a$ and multiply this state from the left to the a-component of p, the other components are unchanged. The reason that we build up protocols from right-to-left is that we view protocols contained in a stack and we follow the convention that the top of a stack is on the left-hand side. It follows from the definition that if $t \in M$ is a trace then $p = tq_0$ denotes a well-defined protocol with $p \in \prod_{x \in X} Z_x^+$. The crucial point is that if $p = tq_0$ is a protocol and $t = at'$ for some $t' \in M$, $a \in X$ then the protocol $t'q_0$ can be computed from tq_0 in constant time by erasing the left most state in the a-component of the protocol tq_0. More generally, if $p = tq_0$ and $t = uv$ then we may compute the protocol vq_0 starting from p in $|u|$-steps. We also shall write $u^{-1}p$ to denote the protocol vq_0 if $p = uvq_0$. This is the same convention as for traces: if $t = uv$ then $u^{-1}t$ denotes the trace v.

If $U = (\prod_{x \in X} Z_x, \delta, q_0, F)$ is a finite asynchronous cellular automata then a protocol $p \in \prod_{x \in X} Z_x^+$ is called *final* if $p = zp'$ for some $z \in \prod_{x \in X} Z_x$, $p' \in \prod_{x \in X} Z_x^*$ with $z \in F$. (We could also view $\prod_{x \in X} Z_x^+$ as a state set of an infinite asynchronous cellular automata where M operates on the left). We are now ready to prove the following result.

Theorem 5.7.1 *There is a construction giving on input a finite noetherian trace replacement system $S \subseteq M \times M$ an algorithm* right_reduce$_\infty$ *which satisfies the following assertions:*

i) *It holds* right_reduce$_\infty(s) \in \mathrm{Irr}(S)$ *for all $s \in M$ and* right_reduce$_\infty$ *terminates in time $O(d_S^2)$.*

ii) *For some systems $S \subseteq M \times M$ the worst-case behavior of* right_reduce$_\infty$ *is $\Theta(d_S^2)$.*

iii) *Whenever $S \subseteq M \times M$ satisfies $G_k(S)$ for some $k \geq 0$ then* right_reduce$_\infty$ *terminates in time $O(d_S)$.*

Proof: For $S \subseteq M \times M$ let $U = (\prod_{x \in X} Z_x, \delta, q_0, F)$ be a finite asynchronous cellular automaton which recognizes the set of reducible traces if they are read from right-to-left. A protocol means an element $p \in \prod_{x \in X} Z_x^+$ as defined above. Define the algorithm right_reduce$_\infty$ as follows

```
function right_reduce∞ (s:trace):trace
var t:trace:=1;
var p:protocol:=q0;
while s ≠ 1
do choose some a ∈ X which is maximal in s;
    s := sa⁻¹; t := at; p := ap;
    ("time: O(1)")
```

if p is a final protocol
("recall that p is final if and only if t is reducible, time: $O(1)$")
then compute some $u \in M$ of minimal length such that
$t = ulv$ for some $(l, r) \in S$, $v \in M$;
("It is crucial that $|u|$ is minimal and to note that this can be done
in time $O(|u|)$. Note also that we must have $l \in aM$,
hence $v = (ul)^{-1}t \in \mathrm{Irr}(S)$ ")
$s := sur$; $t := (ul)^{-1}t$; $p := (ul)^{-1}p$; ("time: $O(|u|)$")
endif
endwhile
return t
endfunction.

It is easy to verify the correctness of the algorithm, i.e., right_reduce$_\infty(s) \in$
$\mathrm{Irr}(S)$ for all $s \in M$, by the following loop-invariants: st is a descendant of the
input trace, t is irreducible and p is the protocol tq_0. For the time complexity
see the comments above. Two points are important: First, the "if-test" can be
performed in constant time. This means we try to find a left-hand side inside
(the stack) t only if we know that such a left-hand side exists. This was the only
reason to work with asynchronous or asynchronous cellular automata. Second,
the factorization $t = ulv$ for some $u, v \in M$, $(l, r) \in M$ with $|u|$ minimal can be
performed in $O(|u|)$ steps. This can be seen, for example, from the representation
of a trace as a tuple of words. Now, if the system S satisfies $G_k(S)$ for some
$k \geq 0$ then we will have $O(|u|) = O(1)$. This proves iii). Assertion i) is obvious
since we do not enter the then-part of the while-loop if t is an irreducible trace.
Assertion ii) is shown in the following example. □

Example: Let (X, D) be given by the graph $a - b - c - d$. Consider the following
special trace replacement system $S = \{bc \Longrightarrow 1, \ ad \Longrightarrow 1\}$. This system is
confluent and the function d_S is linear. Independently of implementation details
the worst-case behavior of the algorithm right-reduce$_\infty$ above will be $\Theta(n^2)$.

Indeed, consider an input trace of the form $s = (ab)^n c^n d^n$. Of course, s
reduces to the empty trace. But the algorithm right-reduce$_\infty$ will perform $\Theta(n^2)$
times the while-loop. Thus, the time complexity of the algorithm can not be
better than $\Theta(n^2)$ even if the whole while-loop could always be performed in
constant time. □

5.8 Cones and Blocks

The property $G_k(S)$ of a noetherian trace replacement system is global on the
left-hand sides. In particular, we can not check it locally and for large systems

it may become very hard to decide whether $G_k(S)$ holds for some given $k \geq 0$.
Furthermore the property is not stable with respect to subsystems. This means
if we delete rules from the system then it can be destroyed. Deletion of rules
occurs during the normalization process of complete systems which is extremely
important to have small systems. Therefore it is desirable to have systems satisfying the property $G_k(S)$ locally on each rule. Thus, we are interested in systems
such that for some $k \geq 0$ we have for each left-hand side l the following local
property $L_k(l) : X(M \backslash MlM) \subseteq MlM \cup X^{(k)}lM$.

Note that this is exactly the restriction of the property G_k to each one-rule
subsystem. The first observation is that it is now superfluous to consider different
k. Indeed let $l, t \in M = M(X, D)$ be traces, $a \in X$ be a letter such that $l \subseteq at$ is
a subtrace, but l is no subtrace of t. If we can not write $at = lt'$ for some $t' \in M$
then there must be a minimal letter x of t such that a and x are independent
but $xl \neq lx$.

In this case l is not a subtrace of $x^k t$ for any $k \geq 0$ and if we factorize
$ax^k t = ulv$ then we must have $u = x^{k+1}$ and $v = 1$. Thus, if l does not satisfy
$L_0(l)$ then $L_k(l)$ is false for every $k \geq 0$. This fact was already mentioned
in Remark 5.6.3 i). We will see in this section that traces satisfying L_0 have
a nice geometric characterization as cones or blocks and that the confluence of
noetherian systems where every left-hand side is a cone or a block can be decided
on the basic critical pairs.

Definition: Let (X, D) be a finite dependence alphabet.

i) A trace $l \in M(X, D)$ is called a *cone* if every letter $x \in X$ which is
   dependent on any letter of l is dependent on all minimal letters of l.

ii) A trace $l \in M(X, D)$ is called a *block* if every minimal letter of l commutes
    with l and every other letter $x \in X$ which does not commute with l is
    dependent on all letters in l.

It follows from the definition that a non-empty cone has exactly one minimal
element. (In particular, it is a *pyramid* in the sense of [Vie86, Def. 5.9].) In the
following this minimal element is called the *top* of the cone. Every element of
the cone is dependent on its top, therefore cones are connected traces. The top
determines the whole dependence relation since a letter $x \in X$ is independent
of a cone if and only if it is independent of the top. More precisely, let $l \subseteq t$
be a subtrace which is a non-empty cone with top a. Then with respect to t
it holds $\mathrm{pre}(a) = \mathrm{pre}(l)$, $\mathrm{ind}(a) = \mathrm{ind}(l)$ and $\mathrm{suf}(a) = (l \backslash \{a\}) \mathrm{suf}(l)$. Trivial
examples of cones are words in a free monoid. On the other hand, vectors in a
free commutative monoid are typical examples of blocks. A block l may have
several minimal elements, it is never connected unless $l = a^n$ for some $a \in X$,
$n \geq 0$. This follows since a letter $x \in X$ commutes with a trace l if and only

if x and l are independent or x is a letter of the alphabet of l and all other letters of alph(l) are independent of x. In particular, every block l has the form $l = \prod_{a \in F} a^{n_a}$ where $F \subseteq X$ is a subset of pairwise independent letters, and $n_a \geq 1$ for all $a \in F$. Thus, a trace l is both, a cone and a block, if and only if $l = a^n$ for some $a \in X$, $n \geq 0$.

The question whether a trace is a cone (a block respectively) is clearly decidable. But observe that the answer to this question depends on the whole dependence alphabet, not only on the alphabet of the trace.

Example: Let t, l be the traces as in Figure 5.3. Then t is a cone and l is a block over the dependence alphabet A. But neither t is a cone nor l is a block over the dependence alphabet B.

Theorem 5.8.1 *Let (X, D) be a finite dependence alphabet and $l \in M(X, D)$ be a trace. Then the following assertions are equivalent:*

i) For every letter $a \in X$ and every trace $t \in M(X, D)$ where l is a subtrace of at it holds that l is a subtrace of t or it holds $at = lt'$ for some $t' \in M(X, D)$.

ii) For all letters $x \in X$ we have $xl = lx$ or $(x, y) \in D$ for all minimal letters y of l.

iii) The trace l is a cone or a block.

Proof: i) $\Rightarrow$ ii): Let $x \in X$ be a letter which is independent of some minimal letter $a \in \min(l)$. Consider the trace $t = x(l \setminus \{a\})$, then l is no subtrace of t, but l is a subtrace of at. Hence $at = lt'$ for some $t' \in M$. But this implies $x = t'$ and $xl = lx$.

ii) $\Rightarrow$ i): Let $a \in X$, $t \in M$ such that l is a subtrace of at but not of t. Let $l' = l \cap t$ in at, then $al' = l$ and $t = ul'u'$ for some $u \subseteq t$ such that every $x \in u$ is independent of a; hence by ii) we have $xl = lx$ for every $x \in u$. Since $ul = lu$ implies $ul' = l'u$, we obtain $at = aul'u' = al'uu' = luu'$.

ii) $\Rightarrow$ iii): Assume that l is not a cone. Then there exist elements $x \in X$, $a \in l$ and a minimal element $y \in \min(l)$ such that $(a, x) \in D$ and $(x, y) \notin D$. By ii) we obtain $lx = xl$. Since x is not independent of l, we have $x \in \min(l)$. Since $(x, y) \notin D$, there are at least two different minimal letters in l. Applying ii) again, we see that every minimal letter in l commutes with l and that l must be a block.

iii) $\Rightarrow$ ii): If l is a block then ii) follows trivially. Hence, we may assume that l is a non-empty cone. Let $\{a\} = \min(l)$ be the top of the cone and let $x \in X$ be any letter such that x is independent of a. Then x is independent of l, hence $xl = lx$.

The equivalence of i), ii) and iii) is thereby shown. $\square$

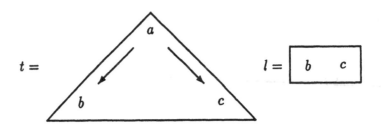

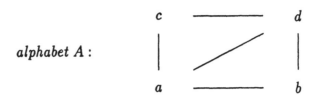

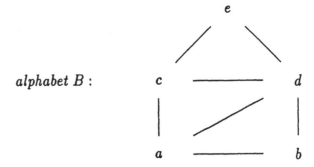

Figure 5.3. A cone t and a block l over A but not over B

Corollary 5.8.2 *Let $S \subseteq M \times M$ be a finite confluent weight-reducing replacement system over a trace monoid M such that every left-hand side is a cone or a block. Then the word problem of the monoid M/S is decidable in linear time.*
□

Remark 5.8.3 i) The previous corollary is a special case of Theorem 5.6.1. Already in this form it is a generalization of [Boo82, Thm. 4.1] since it applies to semi-Thue systems. It generalizes also the result of C. Wrathall [Wra88] on free partially commutative groups.

ii) The well-known result that each finitely generated commutative monoid has word problem decidable in linear time, [Lai67], [FMR68], may be derived from the corollary above as follows: Consider a commuative monoid A on k generators. Let $\gamma_0 : \mathbb{N}^k \to \mathbb{N}$ be the length-function and $\gamma_i : \mathbb{N}^k \to \mathbb{N}$ be the i-th projection, $i = 1, \ldots, k$. The lexical ordering of $\mathbb{N}^k$ is defined by setting $u \succ v$ for $u, v \in \mathbb{N}^k$ if for some $i \in \{0, \ldots, k\}$ it holds $\gamma_j(u) = \gamma_j(v)$ for $j = 0, \ldots, i-1$ and $\gamma_i(u) > \gamma_i(v)$. Using Knuth-Bendix completion, the monoid A is presentable by some finite complete vector replacement system $S \subseteq \mathbb{N}^k \times \mathbb{N}^k$ such that $l \succ r$ for all $(l, r) \in S$, see Corollary 5.1.4, [Buc70], [Gil71] or [BL81]. Since S is finite we effectively find positive integers $a_0 \geq a_1 \cdots \geq a_{k-1} \geq a_k = 1$ such that $\gamma(l) > \gamma(r)$ for all $(l, r) \in S$ where $\gamma : \mathbb{N}^k \to \mathbb{N}$ denotes the weight function $\gamma = a_0 \gamma_0 + \cdots + a_k \gamma_k$. This can be seen by induction on k. The value of a_0 is computed at last. The system S is a weight-reducing trace replacement system and every left-hand side of S is a block. Hence every finitely generated commutative monoid is presentable by such a system. Note that we consider only non-uniform word problems whereas the complexity of the uniform word problem, in which the defining relations as well as the words are regarded as instances of the problem, requires already exponential space for commutative semigroups by a result of [MM82].

We already know from Theorem 5.6.1 that we can decide the confluence of a finite noetherian system if all left-hand sides are cones and blocks. In this special case the decision procedure can be considerably simplified.

We need some preparation. We first consider the situation where a trace t contains two subtraces l_1, l_2 being cones or blocks. We distinguish three cases: 1): both subtraces are cones; 2): one is a cone and the other subtrace is a block; and 3): both are blocks.

Lemma 5.8.4 (Cone-cone-case) *Let $M = M(X, D)$ be a trace monoid and let $l_1 \subseteq t$, $l_2 \subseteq t$ be subtraces of a trace t. Assume that l_1 and l_2 are cones. Then it holds:*

i) If $l_1 \cap l_2 = \emptyset$ then l_1 and l_2 are strictly separated.

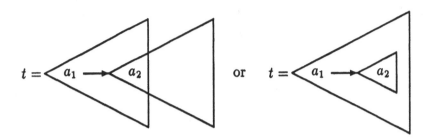

Figure 5.4. Cones with non-trivial intersection

ii) If $l_1 \cap l_2 \neq \emptyset$ then there is no letter between l_1 and l_2; i.e., $l_1 \cup l_2$ is a subtrace of t. This subtrace is a cone and the top of l_1 is contained in l_2 or vice versa.

Proof: We may assume $l_1 \neq \emptyset$ and $l_2 \neq \emptyset$. Let a_1 be the top of l_1 and a_2 be the top of l_2. If $l_1 \cap l_2 = \emptyset$ and l_1, l_2 would be not strictly separated then the tops a_1, a_2 would be not strictly separated, too. But this is clearly impossible. Therefore, we are reduced to prove ii). If $l_1 \cap l_2 \neq \emptyset$ then a_1 and a_2 are dependent, say $a_1 \leq a_2$. If a_2 would not be contained in l_1, then we would have $a_1 < a_2 < b$ for each $b \in l_1 \cap l_2$. This contradicts the fact that l_1 is a subtrace; hence we may assume $a_2 \in l_1$. But then it is clear that there is no letter between l_1 and l_2 and that $l_1 \cup l_2$ is a cone. $\square$

If $t \in M$ is a trace with two subtraces $l_1 \subseteq t$, $l_2 \subseteq t$, $l_1 \cap l_2 \neq \emptyset$ and l_1, l_2 are cones then we may represent this situation by Figure 5.4.

Lemma 5.8.5 (Cone-block-case) *Let $M = M(X, D)$ be a trace monoid and $l_1 \subseteq t$, $l_2 \subseteq t$ be subtraces of a trace $t \in M$. Assume that l_1 is a cone and that l_2 is a block. Then it holds:*

i) If $l_1 \cap l_2 = \emptyset$ then l_1, l_2 are strictly separated.

ii) If $l_1 \cap l_2 \neq \emptyset$ then every letter between l_1 and l_2 is behind l_1 and before l_2. Furthermore every such letter belongs to the alphabet of l_2 and it is not equal to the top of l_1.

Proof: We may assume that l_1 is a non-empty cone with top a.
i) Let $l_1 \cap l_2 = \emptyset$. If l_1, l_2 were not strictly separated then we would have $\mathrm{pre}(l_1) \cap l_2 \neq \emptyset$ and $\mathrm{suf}(l_1) \cap l_2 \neq \emptyset$. But this implies $\mathrm{pre}(a) \cap l_2 \neq \emptyset$ and $\mathrm{suf}(a) \cap l_2 \neq \emptyset$ which contradicts the subtrace property of l_2.

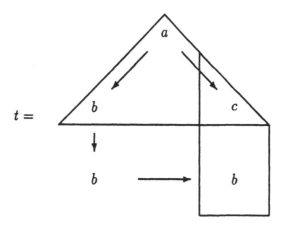

$t =$

Figure 5.5. A cone and a block with non-trivial intersection and a letter between them

ii) Let $b \in l_1 \cap l_2$ and x be a letter between l_1 and l_2. Since x is independent of $b \in l_2$ and there is a path between x and some letter of the block l_2, we have $x \in \mathrm{alph}(l_2)$. A simple observation yields that x depends on the top a of l_1. But then $x \in \mathrm{suf}(a)$ since otherwise there would be a path from x to b. Hence we have $x \in \mathrm{suf}(l_1) \cap \mathrm{pre}(l_2)$. Finally, since a depends on b but x is independent of b, the letters x and a are different.

Example: Let (X, D) be the dependence alphabet $b - a - c$. Then $l_1 = abc$ is a cone but not a block, $l_2 = bc$ is a block. Both traces are subtraces of $t = ab^3c$ with non-trivial intersection (l_2 is a subtrace three times). In one case it happens that there is a letter behind l_1 and before l_2. This is shown in Figure 5.5. $\square$

Lemma 5.8.6 (Block-block-case) *Let $M = M(X, D)$ be a trace monoid and $l_1 \subseteq t$, $l_2 \subseteq t$ be subtraces of a trace $t \in M$ which are not strictly separated blocks. Then $\mathrm{alph}(l_1) \cap \mathrm{alph}(l_2) \neq \emptyset$ and the generated subtrace $\langle l_1 \cup l_2 \rangle$ is a block with $\mathrm{alph}(\langle l_1 \cup l_2 \rangle) = (\mathrm{alph}(l_1) \cup \mathrm{alph}(l_2))$.*

Proof: Let $x, y, z \in t$ such that $x \leq y \leq z$ and $x \in l_1$, $z \in l_2$. If we would have $y \notin (\mathrm{alph}(l_1) \cup \mathrm{alph}(l_2))$ then we would have $l_1 \subseteq \mathrm{pre}(y)$ and $l_2 \subseteq \mathrm{suf}(y)$. Hence, l_1, l_2 would be strictly separated. Since this has been excluded we have $\mathrm{alph}(\langle l_1 \cup l_2 \rangle) = \mathrm{alph}(l_1) \cup \mathrm{alph}(l_2)$. If we would have $\mathrm{alph}(l_1) \cap \mathrm{alph}(l_2) = \emptyset$ then l_1, l_2 would also be strictly separated. Hence, the intersection is not empty and it also follows that $\langle l_1 \cup l_2 \rangle$ constitutes a block in t. $\square$

Example: Let $X = \{a, b, c\}$ where a, b, c are pairwise independent. Let $t = a^3bc^3$ and $l = abc$. We may realize l as a subtrace l_1 of t, for example, by

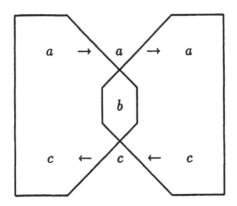

Figure 5.6. Two blocks not strictly separated with non-trivial intersection and letters between them

taking the first a, the only b and the last c of t or as a subtrace l_2 by taking the last a, the only b and the first c of t. In this case the blocks l_1, l_2 have non-trivial intersection, there is a letter before l_1 and behind l_2 and a letter before l_2 and behind l_1. This is shown in Figure 5.6.

We have seen in the lemmata above that there may exist traces $t = \langle l_1 \cup l_2 \rangle$ such that not strictly separated subtraces $l_1 \subseteq t$, $l_2 \subseteq t$ are cones or blocks. It may occur that t contains a letter between $l_1 \subseteq t$ $l_2 \subseteq t$ if one subtrace, say l_2, is not a cone. In all these cases however, $\text{pre}(l_2)$ has a maximal letter which belongs to $\text{alph}(l_2)$ and hence which commutes with l_2. Now, we have seen in Corollary 5.3.10 that these cases can be excluded from the test of confluence. This leads to the following main theorem on cones and blocks.

Theorem 5.8.7 *Let $S \subseteq M \times M$ be a noetherian trace replacement system over $M = M(X, D)$ such that every left-hand side of S is a cone or a block. Then the system is confluent if and only if the following two conditions hold:*

i) The pair (ar, ra) is confluent for all $a \in X$, $(l, r) \in S$ such that $al = la$.

ii) The pair $(r_1 v, u r_2 u')$ is confluent for all $(l_1, r_1), (l_2, r_2) \in S$, u, u', $v \in M$ such that:

1) $l_1 v = u l_2 u'$ and every letter of this trace belongs to l_1 or to l_2,

2) the traces v and u' are independent,

3) $|v| < |l_2|$,

4) no maximal letter of u commutes with l_2,

5) if l_1, l_2 are blocks then we have $u = 1$,

6) if l_1 is a block and l_2 is a cone but not a block then we have $v = 1$ or $u' = 1$.

Proof: Clearly, if S is confluent then i) and ii) hold. For the reverse we may apply Corollary 5.3.10. Thus, by i) it is enough to show the confluence of pairs $(t_1, t_2) = (\text{pre}(l_1)r_1 \text{ind}(l_1) \text{suf}(l_1), \text{pre}(l_2r_2 \text{ind}(l_2) \text{suf}(l_2))$ which result from $t_1 \underset{(l_1,r_1)}{\Longleftarrow} t \underset{(l_2,r_2)}{\Longrightarrow} t_2$ such that $t = \langle l_1 \cup l_2 \rangle$, $l_1 \subseteq t$, $l_2 \subseteq t$ are not strictly separated and no maximal letter of $\text{pre}(l_i)$ commutes with l_i, $i = 1, 2$. We directly see from the three lemmata above that every letter of t belongs to l_1 or l_2. This means that (r_1v, ur_2u') is a basic critical pair.

Now we distinguish several cases:

I) The subtraces $l_1 \subseteq t$, $l_2 \subseteq t$ are blocks. This is the easiest case. It follows from Lemma 5.8.6 that $t = \langle l_1 \cup l_2 \rangle$ is a block of pairwise commuting elements. This implies $\text{pre}(l_1) = \text{pre}(l_2) = 1$. Therefore $t = \langle l_1 \cup l_2 \rangle$ is the uniquely determined minimal trace containing l_1 and l_2. Let $l_1 = \prod_{a \in F_1} a^{m_a}$, $l_2 = \prod_{a \in F_2} a^{n_a}$ then we may write $t = \prod_{a \in F_1 \cup F_2} a^{\max(m_a, n_a)}$. The formula 1) - 6) are obvious from this description.

II) One subtrace is a block the other one is a cone and not a block. Assume first that there is some element before the cone which is not a block. Let l_2 be the cone and l_1 be the block. The top of l_2 is in $l_1 \cap l_2$ because otherwise l_1, l_2 would be strictly separated. It follows $\text{pre}(l_2) = a^k$ for some $k \geq 1$ and $\{a\} = \text{top}(l_2)$. Since l_2 is not a block there is some letter $b \in \text{alph}(l_2)$ with $a \neq b$ and $(a, b) \in D$. Hence there can be no other letter $c \in \text{alph}(l_1)$ with $a \neq c$, because otherwise $(b, c) \in D$, since l_1 is a block and then, $(a, c) \in D$, since l_2 is a cone. But we must have $(a, c) \notin D$ since l_1 is a block. We conclude $l_1 = a^n$ for some $n > k$, $l_2 = a^{n-k}v$ for some $v \in M$ and $t = l_1 v = a^k l_2$. Since $al_2 \neq l_2 a$ the assertions 1) - 6) are easily verified.

Thus, we may assume that there is no element before the cone. Now, switch the notation of the subtraces. Let l_1 be the cone and l_2 be the block. If l_2 is contained in l_1 then the formulae follow with $v = 1$. If l_2 is not contained in l_1 then we have $\text{ind}(l_1) = \emptyset$. We may write $t = l_1 \text{suf}(l_1) = \text{pre}(l_2)l_2 \text{ind}(l_2) \text{suf}(l_2)$. Setting $v = \text{suf}(l_1)$, $u = \text{pre}(l_2)$, $u' = \text{ind}(l_2) \text{suf}(l_2)$ we obtain formulae 1) - 6). A trace t where $\text{suf}(l_1)$, $\text{pre}(l_2)$, $\text{ind}(l_2)$ and $\text{suf}(l_2)$ are all non empty is shown in Figure 5.7.

III) Both subtraces l_1, l_2 are cones without being blocks. By Lemma 5.8.4 we know that $t = \langle l_1 \cup l_2 \rangle = l_1 \cup l_2$ is a cone. The top of t is the top of l_1 or of l_2, say of l_1. Then we have $\text{pre}(l_1) = \text{ind}(l_1) = \emptyset$. We may write $t = l_1 \text{suf}(l_1) = \text{pre}(l_2)l_2 \text{ind}(l_2) \text{suf}(l_2)$. Since $\text{suf}(l_1) \subseteq l_2$ the independence of $\text{suf}(l_1)$ and $\text{suf}(l_2)$ follows. Hence we obtain the formulae 1) - 6) with $v = \text{suf}(l_1)$, $u = \text{pre}(l_1)$, $u' = \text{ind}(l_2) \text{suf}(l_2)$. This concludes the proof. A trace t containing two cones

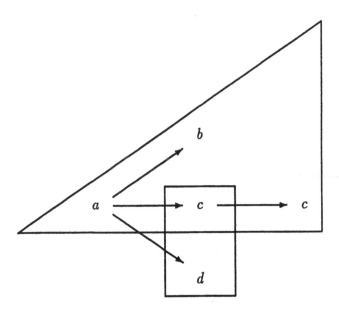

Figure 5.7. A trace containing a cone $l_1 = abc^2$, a block $l_2 = cd$ such that $\operatorname{suf}(l_1) = d$, $\operatorname{pre}(l_2) = a$, $\operatorname{ind}(l_2) = b$, $\operatorname{suf}(l_2) = c$

where $\operatorname{suf}(l_1)$, $\operatorname{pre}(l_1) \operatorname{ind}(l_2)$, and $\operatorname{suf}(l_2)$ are all non-empty is shown in Figure 5.8.
□

Our considerations were based on the question when we can compute irreducible normal forms from right to the left. Of course, there is no mathematical
reason for such a preference. We could also work in the other direction. This
leads to a dual definition of a cone: We say that l is a *reverse cone* if $\operatorname{rev}(l)$ is
a cone, where $\operatorname{rev}(l)$ means the trace which is defined by reversing the ordering
in the labelled partial order of l. Using reverse cones instead of cones we obtain
completely analogous results. In particular, the last theorem has the following
immediate consequence:

Corollary 5.8.8 *Let $S \subseteq M \times M$ be a finite noetherian trace replacement system
such that all left-hand sides are cones (reverse cones respectively) or blocks. Then
it is decidable whether S is confluent on its set of basic critical pairs $BCP(S)$.*
□

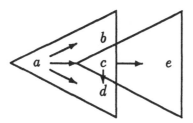

Figure 5.8. A trace containing two cones $l_1 = abcd$, $l_2 = ce$ such that $\text{suf}(l_1) = e$, $\text{pre}(l_2) = a$, $\text{ind}(l_2) = b$, and $\text{suf}(l_2) = d$

Examples:

$$a \; — \; g \; — \; b$$

i) Let (X, D) $=$ $\qquad\qquad\qquad$ $\Big|$ $\qquad\qquad\qquad$, and

$$c$$

$$S = \{gab \Longrightarrow gc,\, abc \Longrightarrow c^2\}$$

Then gab is a cone and abc is a block. According to Theorem 5.8.7 only one critical pair occurs, which is trivially confluent:

$$a \; \longleftarrow \; g \; \longrightarrow \; c$$

$$gc^2 \underset{(gab,gc)}{\Longleftarrow} \qquad \Big\downarrow \qquad \underset{(abc,c^2)}{\Longrightarrow} gc^2$$

$$b$$

Hence S is finite length-reducing confluent system. We may solve the word problem of S in linear time using the function right_reduce$_0$.

ii) Let (X, D) be as above and

$$S = \{ag \Longrightarrow gb,\, bg \Longrightarrow ga,\, cg \Longrightarrow gc,\, g^2 \Longrightarrow 1,\, abc \Longrightarrow 1\}.$$

Then ag, bg, cg, g^2 are reverse cones and abc is a block. It is easily checked that S is noetherian. By Corollary 5.8.8 we see that all critical pairs of S are confluent. Hence S is a finite complete system. It is a symmetric version of an example which was used earlier in the section of the completion procedure on traces. Since Condition $G_0(S)$ holds, right_reduce$_0(s)$ is irreducible for all traces $s \in M(X, D)$.

Although S is not weight-reducing for any weight function $\gamma : X \to \mathbb{N}$, a simple reflection yields that right_reduce$_0$ computes irreducible normal forms in linear time. It is remarkable that an analogous function left_reduce, which works from the left to the right, needs square time in this case.

iii) The example above is a special case of the following group theoretical situation: Let G be a finitely generated group with an abelian normal subgroup A such that the quotient group $E = G/A$ is finite. Let $\{a_1, \ldots, a_k\}$ be a set of generators of A and for each $g \in E$ let s_g be a representative in G. Then A has a presentation by some finite complete weight-reducing vector replacement system $S = \mathbb{N}^k \times \mathbb{N}^k$ and E has a presentation by its multiplicative table. The relations for G are given by the system S and additional equations of the form:

$$
\begin{array}{llll}
s_g s_h &=& s_{gh}\delta(g, h) & \quad, g, h \in E, \quad \delta(g, h) \in A, \\
a s_g &=& s_g a^g & \quad, g \in E, \quad a, a^g \in A.
\end{array}
$$

This leads to a finite trace replacement system over the free partially commutative monoid $M = \mathbb{N}^k * \{s_g; g \in E\}^*$ as follows:

$$
\begin{array}{lll}
T &=& S \cup \{s_g s_h \Rightarrow s_{gh}\delta(g, h); \; g, h \in E\} \\
& & \cup \{a_i s_g \Rightarrow s_g a_i^g; \; g \in E, \; i = 1, \ldots, k\}.
\end{array}
$$

The system T is complete and it defines the group G. The left-hand sides of T are blocks if they belong to S and reverse cones otherwise. The function right_reduce$_0$ computes irreducible normal forms in linear time. Hence the word problem of G is solved in linear time using the system T. Note that we can not expect to find a finite complete semi-Thue system over the set of generators $\{a_1, \ldots, a_k\} \cup \{s_g; g \in E\}$ which defines G. This fails already in the case where E is trivial. Thus the use of traces is essential here.

It is open whether one can join cones and reverse cones into a common concept. But this might be difficult. May be it is impossible. Some problems may be seen in Figure 5.9. In this picture $l_1 = gbc$ denotes a cone, $l_2 = bcg$ is a reverse cone, l_1 and l_2 have non-trivial intersection, but there is a letter between them. We conclude our investigation on traces with this picture. It stands for one of the basic phenomena we met in this theory.

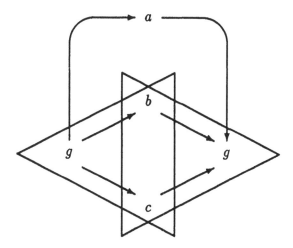

Figure 5.9. A trace containing a cone *gbc* a reverse cone *bcg* and with a letter *a* between them

Conclusion and Outlook

There are basic requirements which should be satisfied by a reasonable model of parallelism. The model should be general enough to cover a large variety of aspects of concurrency. It should be semantically faithful in the sense that essential features are retained. Furthermore, it has to be simple enough such that we are able to understand it; and it must allow efficient computations.

Since these requirements are in general contradictory, we have to compromise. In these notes we have studied Mazurkiewicz-traces, i.e., elements of free partially commutative monoids, as models of parallelism. A trace describes the inherent dependence structure of a concurrent process. Since traces are graphs, they allow to visualize the behaviour of a process immediately. On the other hand, the embedding of traces in a simple algebraic structure leads to efficient algorithms.

Many problems of parallelism can be stated and investigated in terms of traces. Even problems of deadlock and fairness can be studied when infinite traces are used, see [FR85]. The development of further interrelations between concurrency and trace theory is a challenging program.

Our contribution has been mainly concerned with that part of the program which is related to net theory and replacement systems, proving various new results in this area.

The application to net theory has been based on the introduction of local morphisms between Petri nets. We showed thereby that traces are a helpful tool for Petri nets in general and that, from an operational viewpoint, their use is not limited to safe nets.

Local morphisms help to build up Petri net languages in a modular way. Future research should extend this approach. For example: is it possible to apply it to the reachability problem of Petri nets? The decidability of that problem has been shown by E. Mayr, see [May84]. But up to now, the known proofs are rather involved, see also [Kos84], [Reu89]. It would be interesting to know whether our techniques could yield a simplified proof.

For that we probably need a better insight into synchronizations of trace languages. Here, we have settled the question when the synchronization has a local description. We have seen that the answer to that question is *Co-NP-complete*, but in restricted cases we have tractability.

Since a static description of concurrent processes is insufficient, we have introduced replacement systems over traces. This has increased the flexibility enormously. Trace replacement systems provide us with an abstract calculus for transformations on concurrent processes. We hopefully think that this part of the theory will lead us directly to practical applications.

However, because of the novelty of the theory of trace replacement systems many important and interesting questions are still open. We paid the main attention to complete systems and to critical pairs. We have given some sufficient conditions for a finite system to have a finite set of critical pairs, and we have given an upper bound for their number under that condition.

Weaker conditions would be desirable; and the calculation of an upper bound should be improved by a more sophisticated analysis. Both is of high interest for an effective use of the Knuth-Bendix completion over traces.

The Knuth-Bendix procedure is a method for the construction of complete systems. The interest in these systems results from the solvability of their word problem. In order to have really efficient solutions one has to put further restrictions on complete systems. Such a restriction is a weight reduction. We have seen that a finite confluent weight reducing trace replacement system has a word problem solvable in square time. On the other hand, the restriction to semi-Thue systems or vector replacement systems allows linear time. We have extended the linear time results to systems which satisfy a certain global condition on their left-hand sides. This approach includes systems where all left-hand sides are cones and blocks. It is open whether this limitation can be weakened substantially.

Another open question concerns decidability of confluence. It is shown in [NO88] that the confluence of lenght-reducing trace replacement systems is recursively unsolvable. But we do not know whether it is solvable for monadic or special systems, which seems to be a difficult question.

Besides its connection to parallelism, trace theory is also of interest from a purely mathematical viewpoint. Many identities in combinatorics are based on Möbius inversion. Möbius functions for free partially commutative monoids have been studied since the very beginning of that theory, see [CF69]. Here we have settled a conjecture on Möbius functions with the help of complete semi-Thue systems thereby establishing a new bridge between formal power series and semi-Thue systems. Since both, formal power series and semi-Thue systems, are active areas of research in computer science this bridge is very promising and it should be investigated in more detail.

This shows that the current developments in the theory of traces lead into various directions. In fact, the wide range of the theory made it impossible to cover all aspects. For example we have not treated semi-commutation, infinite traces, probabilistic analysis, context free trace languages, connections between recognizability and monadic second order logic, etc. Some open problems in these areas can be found in [Die90b] where a list of research topics has been collected from several authors on occasion of an international workshop organized by the

ESPRIT Basic Research Action No. 3166 ASMICS, [Die90c].

The different perspectives of free partially commutative monoids make trace theory very attractive. We have no doubt that traces will be of continuous interest and importance.

Bibliography

[AH87]   I.J. Aalbersberg and H.J. Hoogeboom. Decision problems for regular trace languages. In Th. Ottmann, editor, *Proceedings of the 14th International Colloquium on Automata, Languages and Programming (ICALP'87), Karlsruhe (FRG) 1987*, Lecture Notes in Computer Science 267, pages 250–259. Springer, Berlin-Heidelberg-New York, 1987.

[AR86]   J. Aalbersberg and G. Rozenberg. Theory of traces. Report 86-16, University of Leiden (The Netherlands), Vakgroep Informatica, 1986.

[AR88]   I.J. Aalbersberg and G. Rozenberg. Theory of traces. *Theoret. Comput. Sci.*, 60:1–82, 1988.

[AW86]   I.J. Aalbersberg and E. Welzl. Trace languages defined by regular string languages. *R.A.I.R.O.-Informatique Théorique et Applications*, 20:103–119, 1986.

[BBMS81] A. Bertoni, A. Brambilla, G. Mauri, and N. Sabadini. An application of the theory of free partially commutative monoids: asymptotic densities of trace languages. In J. Gruska et al., editors, *Proceedings of the 10th Symposium on Mathematical Foundations of Computer Science (MFCS'81), Strbske' Pleso (CSSR) 1981*, Lecture Notes in Computer Science 118, pages 205–215. Springer, Berlin-Heidelberg-New York, 1981.

[BD88]   L. Bachmair and N. Dershowitz. Critical pair criteria for completion. *J. Symbolic Computation*, 6:1–18, 1988.

[BL81]   A.M. Ballantyne and D.S. Lankford. New decision algorithms for finitely presented commutative semigroups. *Comput. and Maths. with Appls.*, 7:159–165, 1981.

[BL87]   R. Book and H.-N. Liu. Rewriting systems and word problems in a free partially commutative monoid. *Inform. Proc. Letters*, 26:29–32, 1987.

[BO84]   G. Bauer and F. Otto. Finite complete rewriting systems and the complexity of the word problem. *Acta Inf.*, 21:521–540, 1984.

[Boo82] R. Book. Confluent and other types of Thue systems. *J. Assoc. Comput. Mach.*, 29:171–182, 1982.

158   Bibliography

[Bra76]   W. Brauer. W-automata and their languages. In A. Mazurkiewicz, editor, *Proceedings of the 5th Symposium on Mathematical Foundations of Computer Science (MFCS'76), Gdansk (Poland) 1976*, Lecture Notes in Computer Science 45, pages 12–22. Springer, Berlin-Heidelberg-New York, 1976.

[Buc70]   B. Buchberger. An algorithm for finding basis a for the residue class ring of a zero dimensional poynomial ideal. *aequ. math.*, 4:374–383, 1970. (Ph.D. Thesis Univ. Innsbruck 1965).

[Buc85]   B. Buchberger. Basic features and development of the critical-pair/completion procedure. In J.-P. Jouannaud, editor, *Proceedings of Rewriting Techniques and Applications (RTA'85), Dijon (France) 1985*, Lecture Notes in Computer Science 202, pages 1–45. Springer, Berlin-Heidelberg-New York, 1985.

[Cei58]   G.S. Ceitin. An associative calculus with an insoluble problem of equivalence. *Trudy Mat. Inst. Steklov*, 52:172–189, 1958. (in Russian).

[CF69]   P. Cartier and D. Foata. *Problèmes combinatoires de commutation et réarrangements*. Lecture Notes in Mathematics 85. Springer, Berlin-Heidelberg-New York, 1969.

[Cho86]   C. Choffrut. Free partially commutative monoids. Technical report 86/20, LITP Université de Paris 7, 1986.

[CL85]   M. Clerbout and M. Latteux. Partial commutations and faithful rational transductions. *Theoret. Comp. Sci.*, 35:241–254, 1985.

[CM85]   R. Cori and Y. Métivier. Recognizable subsets of some partially abelian monoids. *Theoret. Comp. Sci.*, 35:241–254, 1985.

[CM87]   R. Cori and Y. Métivier. Approximation d' une trace, automates asynchrones et ordre des evenement dans les systemes repartis. Technical Report 1-8708, UER de Mathematiques et d' Informatique, Université de Bordeaux I, 1987.

[CP85]   R. Cori and D. Perrin. Automates et commutations partielles. *R.A.I.R.O.-Informatique Théorique et Applications*, 19:21–32, 1985.

[Dic13]   L.E. Dickson. Finiteness of the odd perfect and primitive abundant numbers with n distinct prime factors. *Amer. J. Math.*, 35:413–422, 1913.

[Die86a]   V. Diekert. Commutative monoids have complete presentations by free (non-commutative) monoids. *Theoret. Comp. Sci.*, 46:319–327, 1986.

[Die86b] V. Diekert. Complete semi-Thue systems for abelian groups. *Theoret. Comp. Sci.*, 44:199–208, 1986.

[Die87a] V. Diekert. On the Knuth-Bendix completion for concurrent processes. In Th. Ottmann, editor, *Proceedings of the 14th International Colloquium on Automata Languages and Programming (ICALP'87), Karlsruhe (FRG) 1987*, Lecture Notes in Computer Science 267, pages 42–53. Springer, Berlin-Heidelberg-New York, 1987. Appeared also in a revised version in *Theoret. Comp. Science*, 66:117-136, 1989.

[Die87b] V. Diekert. Some properties of weight-reducing presentations. In: Two contributions to the theory of finite replacement systems. Report TUM-I8710, Technical University Munich, 1987.

[Die89] V. Diekert. Word problems over traces which are solvable in linear time. In B. Monien et al., editors, *Proceedings of the 6th Annual Symposium on Theoretical Aspects of Computer Science (STACS'89), Paderborn (FRG) 1989*, Lecture Notes in Computer Science 349, pages 168–180. Springer, Berlin-Heidelberg-New York, 1989. To appear in revised version in *Theoret. Comp. Science*.

[Die90a] V. Diekert. Combinatorial rewriting on traces. In C. Choffrut et al., editors, *Proceedings of the 7th Annual Symposium on Theoretical Aspects of Computer Science (STACS'90), Rouen (France) 1990*, Lecture Notes in Computer Science 415, pages 138–151. Springer, Berlin-Heidelberg-New York, 1990.

[Die90b] V. Diekert. Research topics in the theory of free partially commutative monoids. *Bull. of the European Association for Theoretical Computer Science (EATCS)*, 40:479–491, Feb 1990.

[Die90c] V. Diekert, editor. *Free Partially Commutative Monoids*. Proceedings of a workshop of the ESPRIT Basic Research Action No 3166: Algebraic and Syntactic Methods in Computer Science (ASMICS), Kochel am See, Bavaria, FRG (1989). Report TUM-I9002, Technical University Munich, 1990.

[Dub86] C. Duboc. Commutations dans les monoides libres: un cadre théorique pour létude du parallelisme. Thèse, Faculté des Sciences de l'Université de Rouen, 1986.

[DV88] V. Diekert and W. Vogler. Local checking of trace synchronizability. In M. Chytil et al., editors, *Proceedings of the 13th Symposium on Mathematical Foundations of Computer Science (MFCS'88), Carlsbad*

(CSSR) 1988, Lecture Notes in Computer Science 324, pages 271–279. Springer, Berlin-Heidelberg-New York, 1988. Revised and extended version in *Math. Syst. Theory* 22: 161-175, 1989: On the synchronization of traces.

[Eil74]   S. Eilenberg. *Automata, Languages and Machines*, volume I. Academic Press, New York and London, 1974.

[ES69]   S. Eilenberg and M.P. Schützenberger. Rational sets in commutative monoids. *Journal of Algebra*, 13:173–191, 1969.

[Fis86]   W. Fischer. Über erkennbare und rationale Mengen in freien partiell kommutativen Monoiden. Report FBI-HH-B-121/86, Fachbereich Informatik der Universität Hamburg, Hamburg, 1986. (Diplomarbeit 1985).

[Fli74]   M. Fliess. Matrices de Hankel. *J. Math. Pures et Appl.*, 53:197–224, 1974.

[FMR68] P.C. Fischer, A.R. Meyer, and A.L. Rosenberg. Counter machines and counter languages. *Math. Syst. Theory*, 2:265–283, 1968.

[FR82]   M.P. Flé and G. Roucairol. On the serializability of iterated transactions. In *Proceedings ACM SIGACT-SIGOPS, Symposium on Principles of Distr. Comp., Ottawa (Canada) 1982*, pages 194–200, 1982.

[FR85]   M.P. Flé and G. Roucairol. Maximal serializability of iterated transactions. *Theoret. Comp. Sci.*, 38:1–16, 1985.

[Gil71]   R. Gilman. Presentation of groups and monoids. *Journal of Algebra*, 57:544–554, 1971.

[GJ79]   M.R. Garey and D.S. Johnson. *Computers and Intractability*. Bell Laboratories, Murray Hill, New Jersey, 1979.

[Gol86]   M.C. Golumbic. *Algorithmic Graph Theory and Perfect Graphs*. Academic Press, New York, 1986.

[HL78]   G. Huet and D.S. Lankford. On the uniform halting problem for term rewriting systems. Lab. Rep. no 28, INRIA, Le Chesnay, 1978.

[Hoa85]   C.A.R. Hoare. *Communicating Sequential Processes*. Prentice-Hall International, London, 1985.

[Hue80]   G. Huet. Confluent reduction: Abstract properties and applications to term rewriting systems. *J. Assoc. Comput. Mach.*, 27:797–821, 1980.

[Jan88]   M. Jantzen. *Confluent String Rewriting*. EATCS Monographs on Theoretical Computer Science 14. Springer, Berlin-Heidelberg-New York, 1988.

[Jou83]   J.P. Jouannaud. Confluent and coherent equational term rewriting systems applications to proofs in abstract data types. In Ausiello G. et al., editors, *Proceeding of the conference of Trees in Algebra and Programming (CAAP'83)*, Lecture Notes in Computer Science 159, pages 269–283. Springer, Berlin-Heidelberg-New York, 1983.

[Kel73]   R. Keller. Parallel program schemata and maximal parallelism I. Fundamental results. *J. Assoc. Comput. Mach.*, 20:514–537, 1973.

[KMN88] D. Kapur, D. Musser, and P. Narendran. Only prime superposition need be considered in the Knuth-Bendix completion procedure. *J. Symbolic Computation*, 6:19–36, 1988.

[Kos84]   S.R. Kosaraju. Decidability of reachability in vector addition systems. In *Proceedings 14th Ann. ACM STOC*, pages 267–281, 1984.

[KS86]   W. Kuich and A. Salomaa. *Semirings, Automata, Languages*. EATCS Monographs on Theoretical Computer Science 5. Springer, Berlin-Heidelberg-New York, 1986.

[Lai67]   R. Laing. Realization and complexity of commutative events. Technical Report TR 03 105-48-T, University of Michigan, 1967.

[Lal79]   G. Lallement. *Semigroups and Combinatorial Applications*. John Wiley & Sons, New York, 1979.

[Lev44]   F.W. Levi. On semigroups. *Bull. Calcutta Math. Soc.*, 36:141–146, 1944.

[LTS79]   P.E. Lauer, P.R. Torrigiani, and M.W. Shields. COSY: A system specification language based on paths and processes. *Acta Inf.*, 12:109–158, 1979.

[Mac70]   S. MacLane. *Categories. For the Working Mathematician*. Graduate Texts in Mathematics 5. Springer, Berlin-Heidelberg-New York, 1970.

[Mar47]   A. Markov. On the impossibility of certain algorithms in the theory of associative systems. *Dokl. Akad. Nauk SSSR*, I,II(55,58):587–590,353–356, 1947. (in Russian).

[May84]   E. Mayr. An algorithm for the general Petri net reachability problem. *Siam J. Comput.*, 13:441–459, 1984.

[Maz77] A. Mazurkiewicz. Concurrent program schemes and their interpretations. DAIMI Rep. PB 78, Aarhus University, Aarhus, 1977.

[Maz87] A. Mazurkiewicz. Trace theory. In W. Brauer et al., editors, *Petri Nets, Applications and Relationship to other Models of Concurrency*, Lecture Notes in Computer Science 255, pages 279–324. Springer, Berlin-Heidelberg-New York, 1987.

[Mét86] Y. Métivier. Une condition suffisante de reconnaissabilité dans un monoïde partiellement commutatif. *R.A.I.R.O.-Informatique Théorique et Applications*, 20:121–127, 1986.

[Mil80] R. Milner. *A Calculus of Communicating Systems*. Lecture Notes in Computer Science 92. Springer, Berlin-Heidelberg-New York, 1980.

[MM82] E.W. Mayr and A.R. Meyer. The complexity of the word problems for commutative semigroups and polynomial ideals. *Advances in Math.*, 46:305–329, 1982.

[MO88] Y. Métivier and E. Ochmanski. On lexicographic semi-commutations. *Inform. Proc. Letters*, 26:55–59, 1987/88.

[New42] M.H.A. Newman. On theories with a combinatorial definition of "equivalence". *Annals of Math.*, 43:223–243, 1942.

[NO88] P. Narendran and F. Otto. Preperfectness is undecidable for Thue systems containing only length-reducing rules and a single commutation rule. *Inform. Proc. Letters*, 29:125–130, 1988.

[Och85] E. Ochmanski. Regular behaviour of concurrent systems. *Bull. of the European Association for Theoretical Computer Science (EATCS)*, 27:56–67, Oct 1985.

[O'D83] C. O'Dunlaing. Undecidable questions related to Church-Rosser Thue systems. *Theoret. Comput. Sci.*, 23:339–345, 1983.

[Ott87] F. Otto. Finite canonical rewriting systems for congruences generated by concurrency relations. *Math. Syst. Theory*, 20:253–260, 1987.

[Ott89] F. Otto. On deciding confluence of finite string rewriting systems modulo partial commutativity. *Theoret. Comput. Sci.*, 67:19–36, 1989.

[Per84] D. Perrin. Words over a partially commutative alphabet. Report no. 84-59, LITP Université de Paris VII, 1984. Also appeared in A. Apostolico, editor, *Combinatorial Algorithms on Words*, Springer NATO-ASI Series, Vol. F12, p.329-340, 1986.

[Per89] D. Perrin. Partial commutations. In *Proceedings of the 16th International Colloquium on Automata, Languages and Programming (ICALP'89), Stresa (Italy) 1989*, Lecture Notes in Computer Science 372, pages 637–651. Springer, Berlin-Heidelberg-New York, 1989.

[Pos47] E. Post. Recursive unsolvability of a problem of Thue. *J. Symb. Logic*, 12(1):1–11, 1947.

[Red63] L. Redei. *Theorie der endlich erzeugbaren kommutativen Halbgruppen.* Number 41 in Hamb. Math. Einzelschr. Physica-Verlag, 1963.

[Reu89] C. Reutenauer. *Aspects mathématiques des réseaux de Petri.* Masson, Paris, 1989.

[Sak87] J. Sakarovitch. On regular trace languages. *Theoret. Comput. Sci.*, 52:59–75, 1987.

[Shi79] M.W. Shields. Adequate path expressions. In G. Kahn, editor, *Proceedings Semantics of Concurrent Computation, Evian (France) 1979*, Lecture Notes in Computer Science 70, pages 249–265. Springer, Berlin-Heidelberg-New York, 1979.

[Squ87] C. Squier. Word problems and a homological finiteness condition for monoids. *Journ. of Pure and Appl. Algebra*, 49:201–218, 1987.

[Vie86] X.G. Viennot. Heaps of pieces I: Basic definitions and combinatorial lemmas. In G. Labelle et al., editors, *Proceedings Combinatoire énumerative, Montreal, Quebec (Canada) 1985*, Lecture Notes in Mathematics 1234, pages 321–350. Springer, Berlin-Heidelberg-New York, 1986.

[WB83] F. Winkler and B. Buchberger. A criterion for eleminating unnecessary reductions in the Knuth-Bendix algorithm. In *Proceedings Coll. on Algebra, Combinatorics and Logic in Computer Science, Györ (Hungary)*, 1983.

[Wra88] C. Wrathall. The word problem for free partially commutative groups. *J. Symbolic Computation*, 6:99–104, 1988.

[Zie87] W. Zielonka. Notes on finite asynchronous automata. *R.A.I.R.O.-Informatique Théorique et Applications*, 21:99–135, 1987.

[Zie89] W. Zielonka. Safe executions of recognizable trace languages by asynchronous automata. In A. R. Mayer et al., editors, *Proceedings Symposium on Logical Foundations of Computer Science, Logic at Botik '89*,

Pereslavl-Zalessky (USSR) 1989, Lecture Notes in Computer Science 363, pages 278–289. Springer, Berlin-Heidelberg-New York, 1989.

Index

This series reports new developments in computer science research and teaching – quickly, informally and at a high level. The type of material considered for publication includes preliminary drafts of original papers and monographs, technical reports of high quality and broad interest, advanced level lectures, reports of meetings, provided they are of exceptional interest and focused on a single topic. The timeliness of a manuscript is more important than its form which may be unfinished or tentative. If possible, a subject index should be included. Publication of Lecture Notes is intended as a service to the international computer science community, in that a commercial publisher, Springer-Verlag, can offer a wide distribution of documents which would otherwise have a restricted readership. Once published and copyrighted, they can be documented in the scientific literature.

Manuscripts

Manuscripts should be no less than 100 and preferably no more than 500 pages in length.
They are reproduced by a photographic process and therefore must be typed with extreme care. Symbols not on the typewriter should be inserted by hand in indelible black ink. Corrections to the typescript should be made by pasting in the new text or painting out errors with white correction fluid. Authors receive 75 free copies and are free to use the material in other publications. The typescript is reduced slightly in size during reproduction; best results will not be obtained unless the text on any one page is kept within the overall limit of 18 x 26.5 cm (7 x 10½ inches). On request, the publisher will supply special paper with the typing area outlined.
Manuscripts should be sent to Prof. G. Goos, GMD Forschungsstelle an der Universität Karlsruhe, Haid- und Neu-Str. 7, 7500 Karlsruhe 1, Germany, Prof. J. Hartmanis, Cornell University, Dept. of Computer Science, Ithaca, NY/USA 14850, or directly to Springer-Verlag Heidelberg.

Springer-Verlag, Heidelberger Platz 3, D-1000 Berlin 33
Springer-Verlag, Tiergartenstraße 17, D-6900 Heidelberg 1
Springer-Verlag, 175 Fifth Avenue, New York, NY 10010/USA
Springer-Verlag, 37-3, Hongo 3-chome, Bunkyo-ku, Tokyo 113, Japan

ISBN 3-540-53031-2
ISBN 0-387-53031-2